Taken for a Ride

How does public transport work in an African city under neoliberalism? Who has the power to influence its changing shape over time? What does it mean to be a precarious and informal worker in the private minibuses that provide public transport in Dar es Salaam? These are the main questions that inform this in-depth case study of Dar es Salaam's public transport system over more than forty years.

The growth of cities and informal economies are two central manifestations of globalization in the developing world. *Taken for a Ride* addresses both, drawing on long-term fieldwork in Dar es Salaam (Tanzania) and charting its public transport system's journey from public to private provision. This new addition to the Critical Frontiers of Theory, Research and Policy in International Development Studies series investigates this shift alongside the increasing deregulation of the sector and the resulting chaotic modality of public transport. It reviews state attempts to regain control over public transport and documents how informal wage relations prevailed in the sector. The changing political attitude of workers towards employers and the state is investigated: from an initial incapacity to respond to exploitation, to the political organization and unionization which won workers concessions on labour rights. A longitudinal study of workers throws light on patterns of occupational mobility in the sector. The book ends with an analysis of the political and economic interests that shaped the introduction of Bus Rapid Transit in Dar es Salaam, and local resistance to it.

Taken for a Ride is an interdisciplinary political economy of public transport, exposing the limitations of market fundamentalist and postcolonial approaches to the study of economic informality, the urban experience in developing countries, and their failure to locate the agency of the urban poor within their economic and political structures. It is both a contribution to and a call for the contextualised study of neoliberalism.

Matteo Rizzo, Senior Lecturer, Department of Development Studies, SOAS, University of London.

Praise for *Taken for a Ride: Grounding Neoliberalism, Precarious Labour, and Public Transport in an African Metropolis*

'To appreciate Matteo Rizzo's *Taken for a Ride* one must first understand what he has done methodologically. He takes public transport in urban Tanzania and embeds this in a series of relationships with other elements of Tanzanian politics, economics and society. Thus we find that public transport has to be understood in its connection not just with Tanzania's equally celebrated and maligned socialist policies dating from the 1960s, but also to the IMF-structural adjustment policies of the 1980s and the changing international landscape of the second-hand automobile trade. We get the views of politicians as well as of drivers and the passengers, the ultimate end-users. Each element is animated and rendered a part of a larger whole so that by the end of it all we get the picture of a living history of urban Tanzania, replete with highs, lows, and many contradictions. The research is astonishing in its range, the writing vivid and clear, and the end result is an insightful and superb contribution to African and Global South urban studies.'

> **Ato Quayson**, New York University, author of Oxford Street, Accra: City Life and the Itineraries of Transnationalism

'Rizzo paints a graphic picture of the insecurity and exploitation of those working on the privately owned buses on which most of the population of Dar es Salaam depend for their local transport. He then shows how international capital—World Bank money allied to the manufacturers of buses, with support from the Tanzanian elite—developed a new monopoly system of buses running on dedicated routes.

Any study that describes developments in public transport in a rapidly growing African city is valuable, but this also breaks new ground theoretically as a study of the political economy of the urban informal sector.'

> **Andrew Coulson**, author of Tanzania: A political economy

'Taken for a Ride is an exciting and innovative contribution to the emerging field of African labour studies. Through substantial field work and a sophisticated critique of market fundamentalism and post colonial theory, Matteo Rizzo brings vividly to life the struggles of public transport workers in the sprawling African coastal city of Dar es Salaam, Tanzania. Carefully avoiding both a romantic optimism and a bland structural pessimism, the author shows how a shared notion of exploitation was constructed amongst the divided transport workers and a new trade union controlled by informal workers was born.'

> **Edward Webster**, Professor Emeritus, University of Witwatersrand

Critical Frontiers of Theory, Research, and Policy in International Development Studies

Series Editors: Andrew Fischer, Uma Kothari, and Giles Mohan

Critical Frontiers of Theory, Research, and Policy in International Development Studies is the official book series of the Development Studies Association of the UK and Ireland (DSA).

The series profiles research monographs that will shape the theory, practice, and teaching of international development for a new generation of scholars, students, and practitioners. The objective is to set high quality standards within the field of development studies to nurture and advance the field, as is the central mandate of the DSA. Critical scholarship is especially encouraged, within the spirit of development studies as an interdisciplinary and applied field, dealing centrally with local, national, and global processes of structural transformation, and associated political, social, and cultural change, as well as critical reflections on achieving social justice. In particular, the series seeks to highlight analyses of historical development experiences as an important methodological and epistemological strength of the field of development studies.

Also in this series

The Aid Lab
Understanding Bangladesh's Unexpected Success
Naomi Hossain

dsa
Development Studies Association

Taken for a Ride

Grounding Neoliberalism, Precarious Labour, and Public Transport in an African Metropolis

Matteo Rizzo

OXFORD
UNIVERSITY PRESS

OXFORD

UNIVERSITY PRESS

Great Clarendon Street, Oxford, OX2 6DP,
United Kingdom

Oxford University Press is a department of the University of Oxford.
It furthers the University's objective of excellence in research, scholarship,
and education by publishing worldwide. Oxford is a registered trade mark of
Oxford University Press in the UK and in certain other countries

First published 2017
First published in paperback 2019

Published in the United States of America by Oxford University Press
198 Madison Avenue, New York, NY 10016, United States of America

British Library Cataloguing in Publication Data

Data available

Library of Congress Cataloging in Publication Data

Data available

ISBN 978–0–19–879424–0 (Hbk.)
ISBN 978–0–19–883905–7 (Pbk.)

To Nic

Acknowledgements

Writing this book has been a pretty long journey. It began back in 1996 when, immediately upon arriving in Dar es Salaam, on my first visit to Tanzania as a Swahili and Political Sciences of Africa student, I developed an uncontainable interest in Dar es Salaam's public transport system and in particular its minibuses. Seemingly ubiquitous, colourful, and full of aphorisms, *daladala*, as these minibuses are known in Swahili, caught my eyes and ears, as a walk anywhere in the city has as its soundscape that of workers calling the stops to attract passengers on board. This fascination was soon accompanied by the obvious reflection that these private buses dominated the transport system of what was once one of the most famous socialist experiments in Africa, prompting my choice to focus my undergraduate dissertation on the city's public transport system, its history, and functioning. The topic was then revisited for my MSc dissertation (in Development Studies at SOAS, University of London). My doctoral work, still at SOAS, but on a different topic and discipline (economic and social history of late colonial and postcolonial development) took me to Dar es Salaam again in 2001–2, where days in the archive would always terminate with visits to a group of *daladala* workers whom I had studied in 1998. When I rejoined academia in 2008, following a three-year career break, my interest in public transport in Dar es Salaam had not yet vanished and so it remained the subject of short research trips in 2009, 2010, 2011, 2013, and 2014. In 2016, twenty years after I first set my eyes on the subject, my research ride ends and this book is going to press! The help, encouragement, and companionship of many people and institutions in Tanzania, the UK, and Italy—the three countries that bind this project together—have helped along the way.

My greatest debt in Dar es Salaam is to the people who found the time to be interviewed and to share the documents on which my research draws. Among my informants, my deepest gratitude goes to the group of *daladala* workers on whom I focused from the word go and throughout the duration of the project. They made a *daladala* station in Dar es Salaam the place where I felt most at home in the city. Memories of chats with them, not only about work in public transport in Dar es Salaam but also about football or how public transport works in London, two of the workers' favourite topics of conversation, and the hilarious banter among them, often carried me through the slow and at times seemingly neverending process of writing. *'Umerudi Matayo! Tunatafuta nini safari hiyo?'*, 'You are back, Matteo! What are we looking for this time around?' I thank them for their hospitality and for their endless willingness to explain their circumstances.

Research for this book has been made possible by the financial support of a number of institutions where I have worked over the years, and from which I have picked up many of the insights that shape the book, in ways of which I am not even fully aware. The African Studies Centre, University of Oxford, where I worked as a lecturer from 2008 to 2010, funded research trips in 2009 and 2010. In 2011, while a Smuts Research Fellow at the Centre for African Studies in Cambridge, I undertook a research trip funded by the Smuts Memorial Fund. Support and encouragement by the then Centre Director, Megan Vaughan, was crucial to elaborate early drafts of the book proposal. A British Academy Small Research Grant (SG101976) funded further rounds of fieldwork in 2013 and 2014, which I undertook while based at SOAS, University of London, which I joined in 2012 and where I currently work. Two small grants from SOAS Faculty of Law and Social Sciences generously funded the work of Danisha Kazi and Kevin Deane, who provided excellent research assistance to this project. Some of the materials on which this book draws have been previously published and revised for inclusion in this book. I thank the following publishers for allowing me to reuse parts of my work: Taylor and Francis, for reuse of parts of a 2013 article in the *Review of African Political Economy* 40 (136): 290–308, and of a 2015 article in *The Journal of Development Studies* 51: 149–61. Oxford University Press allowed me to reuse parts of a 2015 article in *African Affairs* 114 (455): 249–70. Cambridge University Press allowed me to reuse parts of a 2002 article in the *Journal of Modern African Studies* 40 (1): 133–57. John Wiley & Sons allowed the reuse of parts of a 2011 article in *Development and Change* 42 (5): 179–206. Anson Stewart kindly allowed me to reproduce his map of the main *daladala* routes in Dar es Salaam, presented in Figure 2.1.

For someone interested in radical and interdisciplinary political economy, like myself, SOAS provides a remarkably exciting place in which to work. I owe a debt of gratitude to Dave Anderson, Alastair Fraser, Graham Harrison, Deborah Johnston, Claire Mercer, Carlos Oya, Tim Pringle, John Sender, Megan Vaughan, Leandro Vergara-Camus, and Elisa Van Waeyenberge for their careful reading of one or more chapters of the book, and their generous comments, and above all for encouragement at different times of this project. Henry Bernstein, Chris Cramer, and Ben Fine read the whole manuscript, and provided very constructive comments to strengthen its argument and readability. Their commitment to collegiality at times of very heavy workloads is truly inspiring.

The Development Studies Association series editors, Andrew Fischer, Uma Kothari, and Giles Mohan, and, at Oxford University Press, Adam Swallow and the anonymous book referees were extremely supportive throughout the project—always forthcoming with witty suggestions to improve the text by saving the reader from too much detail while at the same time encouraging and appreciating the value of the fine-grained study that this book presents.

In Tanzania I owe thanks to my Swahili teacher at 'L'Orientale in Naples', Abedi Tandika, not only for his inspiring way of passing on his passion for Swahili language but also for facilitating an unforgettable stay at the *Chuo cha Biashara* of Dar es Salaam (the Dar es Salaam Business College) in 1998. There I met Martin Mayao, one of my three roommates, who would later warmly welcome me to live with him, his wife Greta, and his brother David, in 2001–2, by which time he owned two *daladala*. Martin was a great host, and on top of that never failed to find time to explain to me how *daladala* work, how he operated his small fleet, and to point me in the right direction to find out more myself. Brian Cooksey, a veteran researcher on Tanzania, was always a source of connections and useful advice on the latest twists in urban transport matters in Dar es Salaam and, more broadly, on Tanzania. Dario De Nicola and Marina Mazzoni were great hosts at Cefa Hostel, where I have stayed for all my visits from 2009, for both their curiosity and down-to-earth engagement with my research.

Families and friends are really important to those writing books, and I am no exception. Among many, I especially thank Jonny Donovan, Ben Rawlence, and Gemma and Juma Juma for ongoing support and curiosity. Juma was always there when I needed him for translations of Swahili interviews or press cuttings, and to solve the frequent IT problems I seem to run into ('back up Matayo, back up'). My in-laws, Nigel and Lizzie, and my brother- and sisters-in-law, Seb, Alice, and Anna, struck the perfect balance between asking enough but not too often about progress with the book, and for generous support with childcare. *Mens sana in corpore sano*, my ancestors used to say, and they had a point. I owe thanks to Luca Dogliotti, Mark Draper, Paul Herring, Joe Perkins, and Jimmy Callus for the gruelling clashes on the tennis court and the release from work that they provided me. Back in Sicily, a huge network of friends and family, spanning three generations, provided much-needed injections of home warmth during visits and were often a source of very useful questions from a lay person on what the book is about and why it matters. There are too many to thank and the risk of leaving someone out is too high, but you know who you are! I owe a great deal to my parents and my sister Bianca. Above all, I am grateful to them for demonstrating, through their example, the importance of intellectual honesty and courage. Last, but not least, my kids, Luca, Nina, and Jack, and my wife, Nic, were at once the loveliest drive and distraction to finish the book. They put up with ever so slippery deadlines with admirable patience and *almost* endless faith. 'Will you ever finish your book?' These words from Luca, my oldest son, said a couple of years ago while lifting a book on Africa on my desk with clear despondency in his eyes, urged me on big time to finally complete the manuscript. The greatest debt of all is to my wife, for her unfailing love and support. Nic lent her sharp mind as the sounding board of many of the ideas presented in the book. She was always there for me when writing became tough and provided plenty of childcare support to allow fieldwork and far too much 'weekend' writing. I simply could have not written this book without her.

London
July, 2016

Contents

List of Figures   xvii
List of Tables   xix
List of Acronyms   xxi

1. Taken for a Ride: Rethinking Neoliberalism, Precarious Labour,
   and Public Transport from an African Metropolis   1
   1.1 Early Impressions: Urban Public Transport as Functional Chaos   1
   1.2 Structure and Agency in the African City   3
   1.3 Structure and Agency in the African Informal Economy   7
   1.4 Class Matters   12
   1.5 Neoliberalism, Post-Socialism, and Public Transport   14
   1.6 Methodology of the Book and its Chapters   21

2. Public Transport in Dar es Salaam: From State Monopoly to
   Neoliberalism (1970–2015)   27
   2.1 Introduction   27
   2.2 State-Provided Public Transport: 1970–83   28
   2.3 The Privatization and the Progressive Deregulation of
       Public Transport: 1983–2001   31
   2.4 Deregulation, Privatization, and the Quality of Bus Public
       Transport   40
   2.5 Feeble Attempts to Regain Public Control: 1999–2015   45

3. 'Life Is War': Capital and Informal Labour in Bus Public Transport   51
   3.1 Introduction   51
   3.2 Informal Economy as Self-Employment?   53
   3.3 The 2006 Integrated Labour Force Survey: Definitions and
       Patterns of Employment   57
   3.4 From Statistical Fiction to Employment Realities: The Case
       of the *Daladala*   61
   3.5 Bus Owners in Dar es Salaam   63
   3.6 *Daladala* Workers   66
       3.6.1 Juma Masuka   68
       3.6.2 Kudo Boy   69
       3.6.3 Uwazi   69

Contents

3.6.4 Kajembe   69

3.6.5 Rajabu   70

3.6.6 Rama   70

3.6.7 Asenga   71

3.7 The Employment Relationship in *Daladala*   71

3.8 The 2006 ILFS Questionnaire and Informal Wage Employment: Lost in Translation?   76

3.9 Informal Wage Employment: Invisible and yet Central   78

4. The Politics of Labour 1: The Quiescent Period (up to 1997)   81

4.1 The Criminalization of the Workforce   81

4.2 The Sources of Workers' Power   84

4.3 The Spatial Unit of Work   86

4.4 Labour Heterogeneity: The Phenomenology of Transport Workers   86

4.4.1 *Daladalamen* 'with a Livelihood'   87

4.4.2 People 'on the Bench'   87

4.4.3 'Those Who Hit the Tin'   89

4.5 Workers' Associationism: Forms and Limits of Solidarity   90

4.5.1 Managing but not Challenging Precariousness   90

4.5.2 The 'Struggle over Class'   94

4.6 Transport Workers' Horizontal Mobility and its Implications   96

4.7 United they Stood, Divided they Fell   98

5. The Politics of Labour 2: Struggling for Rights at Work (1997–2014)   100

5.1 Informalization and Rights at Work   100

5.2 From Political Quiescence to Political Organization: Early Days, 1995–2000   105

5.3 The Construction of a Shared Meaning of Exploitation   108

5.4 Labour Rights through Collective Bargaining   112

5.5 Barriers to the Enforcement of Employment Contracts   114

5.6 Labour Rights: Bringing the State Back In   115

5.7 A New Political Subject: Trade Unions, the Informal Economy, and Labour Rights   118

5.8 Contextualizing Workers' Power and Realms of Possibility   119

6. Tracing Occupational Mobility/Immobility among Informal Transport Workers   122

6.1 Hitting a Moving Target: Methodological Issues   122

6.2 Histories of Occupational Immobility   129

6.2.1 Juma Masuka   129

6.2.2 Uwazi   130

6.2.3 Kajembe and Ngaika   131

6.2.4 Sulemani   131

6.3 Histories of Occupational Mobility   132

6.3.1 Rajabu		133
6.3.2 Abasi		134
6.3.3 Dotto		135
6.3.4 Asenga		136
6.3.5 Mudi and Kulwa		138
6.4 Workers' Trajectories: Predictable?		139

7. The New Face of Neoliberalism: The Bus Rapid Transit Project
in Tanzania (2002–16)   142

7.1 The Political Economy of BRTs   142
7.2 The BRT Evangelical Society   148
7.3 The Ideology of BRT in Dar: Whose 'Better City for
Better Times'?   152
7.4 Making Sense of Delays in the Implementation of DART   153
7.5 The Deeper Roots of Lack of Government Support   157
7.6 Towards the Implementation and Domestication of BRT:
2014 Onwards   162
7.7 What Can President Magufuli Do?   165
7.8 BRT Tensions as 'Actually Existing Neoliberalism'   170

8. Conclusion: Taken for a Ride   171
8.1 Cities of Ghosts: Bringing People Back In   171
8.2 Grounding Neoliberalism   176

Appendix A: Questionnaire and Summary of Results   181
Appendix B: Labour Mobility, December 2001–June 2002   183
Glossary   185
References   187
Index   207

List of Figures

2.1 Major *daladala* routes in Dar es Salaam, 2011   28

2.2 Bus fares in US$ in Dar es Salaam, 1983–2013   35

2.3 Number of *daladala* in Dar es Salaam, 1983–2010   37

3.1 'Life is War'   52

3.2 *Kazi mbaya; ukiwa nayo* ('Bad job; if you have one!')   67

3.3 'Money Torture'   74

6.1 Asenga and his kiosk   137

8.1 *Na sisi watu. Tutaheshimiana tu* ('We are also humans.
    We will respect each other')   176

List of Tables

2.1 Public buses in service, 1974–98   29

2.2 Bus fares and currency devaluation trends, 1983–91   33

2.3 Trends in registered private buses, 1983–91   33

2.4 Bus fares and currency devaluation trends, 1991–2009   34

2.5 Trends in registered *daladala*, 1991–2010   36

2.6 Trends in student and adult bus fares, 1983–96   43

2.7 Licensed buses (*daladala*) by size, March 2008–June 2009   49

3.1 Employment status by sector of main employment, 2006 (main activities)   58

3.2 Structure of employment by sector, male and female, 2006 (selected subsectors, main activity only)   59

3.3 Structure of employment by sector, male and female, 2006 (selected sectors, secondary activity only)   60

List of Acronyms

AICD	Africa Infrastructure Country Diagnostic
BRT	Bus Rapid Transport
CCM	*Chama cha Mapinduzi* (Party of the Revolution)
CHADEMA	*Chama cha Demokrasia na Maendeleo* (Party for Democracy and Development)
COTWUT	Communication and Transport Workers Union of Tanzania
CSAE	Centre for the Study of African Economics, Oxford
CTLA	Central Transport Licensing Authority
DARCOBOA	Dar es Salaam Commuter Bus Owners Association
DART	Dar Rapid Transit
DCC	Dar es Salaam City Council
DMT	Dar es Salaam Motor Transport Company
DRTLA	Dar es Salaam Transport Licensing Authority
EMBARQ	The WRI Initiative for Sustainable Urban Mobility
ICFTU	International Confederation of Free Trade Unions
IEA	International Energy Agency
IFI	International Financial Institution
ILFS	Integrated Labour Force Survey
ILO	International Labour Organization
IMF	International Monetary Fund
IMTS	Integrated Mass Transit Systems
ITF	International Transport Workers' Federation
ITDP	Institute for Transportation and Development Policy
LRT	Light Rail Transit
JICA	Japan International Cooperation Agency
KAMATA	*Kampuni ya Mabasi ya Taifa* (National Bus Company)
MCT	Ministry of Communication and Transport of Tanzania
MKURABITA	*Mpango wa Kurasimisha Rasilimali na Biashara za Wanyonge Tanzania* (Property and Business Formalization Programme)

MUWADA	*Muungano wa Watu Wanaosafirisha Daladala* (*Daladala* Owners' Association)
NBS	National Bureau of Statistics of Tanzania
NEPAD	New Partnership for Africa's Development
NGO	Non-Governmental Organization
OECD	Organization for Economic Co-operation and Development
PAP	Project-Affected People
PLC	Cordial Transportation Services
PPP	Public–Private Partnership
SAIIA	South African Institute of International Affairs
SAP	Structural Adjustment Programme
SSA	Sub-Saharan Africa
SSATP	Sub-Saharan Africa Transport Policy Program
SUMATRA	Surface and Marine Transport Regulatory Authority
TANESCO	Tanzania Electric Supply Company
TANU	Tanganyika African National Union
TARWOTU	Tanzania Road Transport Workers Union
TAZARA	Tanzania–Zambia Railway
TDA	Tanzania Drivers Association
TRA	Tanzania Revenue Authority
UATP	African Association of Public Transport
UDA	*Shirika la Usafiri Dar es Salaam* (Dar es Salaam Transport Company)
UDA-RT	Usafiri Dar es Salaam Rapid Transit
UN	United Nations
UNDP	United Nations Development Programme
URT	United Republic of Tanzania
UWAMADAR	*Umoja wa Madereva na Makondakta wa Mabasi ya Abiria Dar es Salaam* (Dar es Salaam *Daladala* Drivers' and Conductors' Association)
UWAMATA	*Umoja wa Madereva wa Mabasi Tanzania* (Tanzania Upcountry Bus Drivers' Union)
WEA	Workers' Education Association of Zambia
WRI	World Resource Institute
WTO	World Trade Organization

1

Taken for a Ride

Rethinking Neoliberalism, Precarious Labour, and Public Transport from an African Metropolis

1.1 Early Impressions: Urban Public Transport as Functional Chaos

Utter chaos but somehow functional. This was the first impression I had, on my first visit to Dar es Salaam in 1996, of the city's transport system and of the privately owned minibuses, or *daladala*, as they are called in Swahili, which provide its public transport.[1] Such an impression came from the sight, ubiquitous in the city, of tireless workers frantically cramming passengers into aged buses—some of these vehicles were obviously past retirement date—and they are routinely loaded well beyond the vehicle's carrying capacity. Any passenger seemed welcome on board, except school pupils who owed their unpopularity with bus crews to their entitlement to concessionary fares. As a result, at stations and bus stops, school pupils in uniform would gang up in small groups to persuade conductors, with words and, if necessary, by sheer force, to allow them to board buses.

Decorated with drawings and writing by their workers, the buses act as a canvas for drivers' and conductors' opinions on various themes, including what it means to work on buses in an African metropolis. The drawing on the back of one bus of two wrestling fighters fiercely facing off before a fight, captioned by the writing 'Warning: No ring, no rules, no referee', aptly sums up the trademark cut-throat competition between one *daladala* and another

---

[1] The name *daladala* comes from the Swahili name for the 5 shilling coin, the *dala*, which in turn came from the fact that the coin was worth one dollar when it was first introduced in 1972. As the first fare private operators charged in 1983 was 5 shillings, bus conductors used to ask passengers for *dala, dala*.

and the machismo that often accompanies it. Fuelled by such competition, overloaded buses speed through the city, overtake dangerously, ride through red lights, on pavements, and outside allocated routes in search of passengers. One tragic consequence of these practices has been that far too many Dar es Salaam residents have died, as *daladala* are involved in a huge proportion of lethal car accidents in the city, as high as 93 per cent in 1992 (*Daily News*, 17 May 1993). Nonetheless, it is under such a system that the majority of people in Dar es Salaam, a city that has been rapidly growing for the past six decades and is now home to over four million residents, have travelled around since 1983. From that year, private buses were allowed to provide public transport following the demise of the state-owned public transport company which had until then provided the service under a monopoly regime—a shift Dar es Salaam shared with many cities in developing countries.[2]

This book presents the results of my efforts to deepen this early impression and understand the origins, logics, and tensions of what at first appeared as functional chaos. This entailed a journey through the history, economic organization, and politics of public transport in Dar es Salaam, in its transition from state provision of the service in the 1960s and 1970s to its progressive privatization, liberalization, and informalization from 1983 to the present. Looking at Dar es Salaam through this prism, this book aims to contribute to two thematic literatures: on African cities and their informal economies. Intriguingly, references to chaos, dystopia, and their opposites, order (if unconventional and not Eurocentrically defined) and functionalism, and attempts to reconcile these poles, are very common in the vast literature that exists on both of these themes. In a sense, they define the 'clash of ideas about where African cities are heading today' (Freund 2007: 165; see also Gandy 2005, for a review of both dystopic and hopeful narratives on Lagos). A useful starting point to understand what is at stake in this case study of one transport system in one African city is thus to navigate this literature, and to review its leading voices, rather than aim for an exhaustive coverage of it. My goal is to sketch out the reasons for such a conflicting understanding of the African urban landscape and its informal economies, and the research agendas that both derive from and inform such divergent readings.

---

[2] On Nairobi, see Khayesi (1998); waMungai and Samper (2006); Salon and Gulyani (2010); on Kumasi, see Adarkwa and Tamakloe (2001); on Accra, see IBIS Transport Consultants Ltd (2005); on Abidjan, see Adoléhoumé and Bonnafis (2000); for two overviews with two pages on the urban transport outlook on a wide range of sub-Saharan countries' main cities, see Kumar and Barrett (2008); Kouakou and Fanny (2008).

1.2 Structure and Agency in the African City

The leading example of perspectives on the urban experience in developing countries which place a strong emphasis on dystopia and chaos is work by Mike Davis (2006). Behind the apocalyptic tone of his *Planet of Slums* lies the fact that urbanization in Asia, Africa, and Latin America—unlike urbanization in the nineteenth century in the now-developed world—has not been accompanied and driven by economic growth and industrialization. This defines its historical novelty, and explains the tragic reality of 'pollution, excrement and decay' in which more than one billion people live in the shanty towns of cities in developing countries. The employment situation in such places is as grim as the housing conditions. Davis reminds us that these are 'cities without jobs'. The number of job seekers has grown rapidly, while the capacity of the formal economy to absorb labour has decreased rather than expanded. As a consequence of this imbalance, the majority of people are pushed towards work in the informal economy, the last-resort employer of this 'surplus humanity'.

Davis's apocalyptic approach therefore strongly emphasizes the structural causes that led to this planet of slums without jobs. The lack of industrialization and of employment creation, Davis argues, reflects the demise of the postcolonial state development project in the 1970s and the pernicious consequences of neoliberalism and of structural adjustment programmes that came in its wake from the 1980s onwards. The urban poor have thus been hit twice, as the withdrawal/failure of the public sector in providing key services such as water, health, education, sanitation, and housing has gone hand in hand with the lack of jobs in a labour market 'as densely overcrowded as the slums themselves'. A *daladala* worker seems to nod to Davis, by writing on his bus *Usiku mbu, nzi mchana* ('Mosquitoes at night, flies in the day').

The strength of Davis's contribution, and arguably its intended goal, is the systematic debunking of mainstream fantasies of slum improvements and of the potential for large-scale poverty eradication and economic mobility within the informal economy. His focus on how structural forces affect the urban poor's experience of the city in developing countries is important, if chilling. This very focus on structures, however, has been, for some, excessive. A common criticism of this apocalyptic narrative is its tendency to sensationalism and its lack of attention to the historical agency that the poor might exert, either to change the grim conditions in which they live and work, or even to survive them (Locatelli and Nugent 2009; Satterwhite 2006). As Myers notes, Davis is 'so fixated on exploding slums, with no hope for poverty alleviation, and urbanism...and so driven towards the worst of the worst-case scenarios and "pathologies" that we, the readers, can only abandon hope, and turn tail heading elsewhere' (Myers 2011: 6).[3] Others have argued that more descriptive 'exploding' of the slums is necessary. Lumping the entire population of Soweto, home as it is to about two million residents in Johannesburg, under the category of slum resident, conceals its 'phenomenal diversity, including wealthy suburbs, serviced by modern shopping malls and a golf course, as well as "respectable working class communities"' (Zeilig and Ceruti 2007: 3).

By contrast, hope and a focus on the agency of urban residents are the distinctive features of a burgeoning literature that claims to put forward a way of understanding the African city, including its materiality, without the shortcomings and teleology that they perceive to define work informed by critical, and often pessimistic, political economy. Work by scholars such as Simone (2004a, 2004b, and 2005), Pieterse (2008), and Robinson (2006) has taken issue with the normative and Eurocentric nature of such dystopic narratives on urbanization in the South, in a move away from materialist explanations of urban realities centred on economic failure. In Simone's words, urban life should not be seen as 'a series of policies gone wrong'. On the contrary, agency and 'determination by urban Africans to find their own way' and 'the resourcefulness' and 'astute capacity' on which they draw hold the key to understanding urban society in Africa (Simone 2005: 1; see also Tati 2001). Along similar lines, Pieterse (2008: 100) argues:

> ...our present thinking about meeting the challenge of slums is profoundly impoverished precisely because we locate these places with their teeming complexities in a black box devoid of complex agency and determinacy, which is nonetheless unconsciously ascribed to parts of the modern city that are considered developed or settled...

The implications of such an approach are both theoretical and political/practical. On the former, the goal is the formulation of a better (and postcolonial) urban theory, which is able to reinsert from 'off the map' the urban experiences of Third World cities. 'All cities are best understood as ordinary', advocates Robinson (2006: 1), in stark contrast with the dystopic narrative on Third World cities. At a practical level, what is at stake in opening our eyes to 'the phenomenology and practices of the "everyday"' (Pieterse 2008: 9) or 'the ordinary' and to the capacity displayed by African cities to generate 'a new urban sociality even under dire conditions' (Simone 2004b: 314) is that such

---

[3] Against this reading of Davis's work on slums as hopelessly pessimistic, it is important to note that Davis ended his *Planet of Slums* by laying out the plan to write, with Forrest Hylton, a sequel book on 'the history and future of slum-based resistance to global capitalism' (Davis 2006: 201, 207), on the basis of 'concrete, comparative case studies', exploring 'the bewildering variety of responses to structural neglect and deprivation' and 'the myriad acts of resistance'. This sequel book was never written, for reasons that I do not know. However, the very fact that it was planned raises questions about dismissing Davis's work as blind to the agency of the poor.

practices 'must be the touchstone of radical imaginings and interventions'. As African cities are built on 'people as infrastructure', policy formulation must, we are told, draw on such infrastructure. In doing so, as Pieterse boldly claims, the goal is 'to recuperate the constitutive humanity and, by extension, generative powers of the ordinary' (Pieterse 2008: 10).[4]

There is much to unpack in this narrative. What is the 'ordinary'? What do concepts such as 'people as infrastructure' and 'generative powers' really mean? From what sources do 'generative powers' derive and what do they generate? To what extent are such 'generative powers' able to make a dent on 'the difficulties and brutalities of grinding poverty and persistent terror', that Pieterse (2008: 9) reassures us, he 'is not to romanticize'? And do such contributions have anything to say about how such grinding poverty is generated in the first place? Can one reasonably think of *radical* interventions to improve the lives of people living in poverty without addressing the *root* or *structural* causes of it? Does the 'acknowledgement of the powerful role of economic forces in shaping the urban milieu' necessarily imply 'any sort of teleology in urban history' (Storper and Scott 2016: 10)? Reading such work in search of answers to these questions reveals how its formulation of the proposed alternative remains woefully vague. Take, for example, Simone's notion of 'people as infrastructure' which, he claims, helps us to understand that 'something else besides decay might be happening' (Simone 2004a: 407) in the inner city of Johannesburg. His writing emphasizes the 'open ended, the unpredictable' and 'provisional intersections of residents that operate without clearly delineated notions of how the city is to be inhabited and used'. There is a reference (although unsubstantiated) to the 'intensifying immiseration of African urban populations' (is this not normative?) and that the 'majority of African urban residents have to make what they can out of their bare lives' (Simone 2004a: 428). Yet, one reads, 'this seemingly minimalist offer—bare life—is somehow redeemed' due to its 'innumerable possibilities of combination and interchange that preclude any definitive judgement of efficacy or impossibility'. But can this reference to the endlessly open-ended really stand up to scrutiny?

Writing from the Marxist political economy tradition, in a direct exchange with some of the scholars above, Watts (2005) underlined the significant theoretical and empirical limitations of this kind of work, and the link

---

[4] De Boeck and Plissart's (2004) monograph on Kinshasa differs from such narratives as it is based on a thorough engagement with fieldwork and the empirical, and includes breathtaking photos. However, it shares some important similarities as its message tends to unduly romanticize, without much analysis, the ethics of a mythical rural past on which Kinshasa residents draw to navigate the city. Furthermore, despite a certain hopeful ring to much of its story, the book fails to articulate what the concrete possibilities for a better Kinshasa are, as is visible in the concluding section of the book. (Mis)named the '(im)possibilities of the possible', the conclusion presents the vision of another Kinshasa by two artists and no real basis of hope for 'ordinary' citizens.

between the two. There was, he pointed out, very little empirical data, and no information whatsoever on how such data had been gathered. There was no information, apart from anecdotes, on 'what forms of sociability' were generated and 'with what consequences' (Watts 2005: 184). As this was not spelled out, what transpired was a 'desperate search for human agency (improvisation, incessant convertibility) in the face of a neoliberal grandslam'. Nuttall and Mbembe disagreed, and argued that the lack of evidence reflected the fact that the nature of reality in African cities 'is fundamentally elusive' (Nuttall and Mbembe 2005: 196). They presented, in response, a manifesto on why it is so difficult to move beyond vague formulations:

> The difficulty with having to act through the provisionality that 'people as infrastructure' implies is that the meanings of the tactics employed can hardly be pinned down. What works in these arrangements and under what circumstances is very difficult to disentangle from or extend to the larger realm of everyday operations. Categories such as ethnicity, nationality, gender, age, and employment status all intersect but in ways such that each, taken on its own, loses explanatory power. (Nuttall and Mbembe 2005: 200)

While pinning down how such categories play themselves out on the ground is indeed challenging, work by scholars who have not been so discouraged by the elusive nature of African urban reality is still capable of advancing knowledge in important ways. For example, Alexander and co-writers explore the ways in which class (included under 'employment status' in Nuttall and Mbembe's list) and other markers of difference such as religion and gender regulate daily life in Soweto, Johannesburg (Alexander et al. 2013). It thus seems that Nuttall and Mbembe's methodology, rather than being a consequence of the elusive nature of African urban realities, is a matter of preference and choice, and a choice that is shared by many other scholars writing on urban Africa. It is, however, questionable, as their preference for light or no engagement with the empirical, and the vague meaning of their references to the 'ordinary' and to the 'everyday', with no clear evidence of what people actually do, critically undermines the analytical purchase of such approaches.

Furthermore, it produces extraordinary cities, which resemble, in a sense, cities of ghosts. Their inhabitants, the amorphous 'urban poor', or 'people at the grassroots', tend to float mid-air, unhinged from the material and the economic. It is this incapacity and/or lack of interest in pinning down materiality, the bones of everyday life and social order, and the resulting reliance on shallow concepts such as 'people as infrastructure', that makes implausible the claim that 'radical imaginings and interventions' should, or even can, be built on the social fabric created by the poor. In sum, such approaches appear as the mirror image of the apocalyptic narratives that they criticize. Their emphasis on the agency and functionalism of initiatives by the poor suffers from the lack of any substantive attention to the economic and political structures in which the poor are located, despite claims to the contrary. Pieterse worries that 'There is a real danger that the material deprivation that accompanies slum life becomes the sole lens through which these communities, households and urban actors are understood and engaged' (Pieterse 2008: 32). In contrast, my worry is that the steady flow of romantic and unsubstantiated celebrations of the choices and repertoires of 'people at the grassroots' crowds out an understanding of the concrete realities they face, and thus any possibility of assessing the meaning and impact of their actions, and of reflecting on the triumphs and tragedies that they experience in seeking to cope with and overcome the challenges of their lives.

1.3 Structure and Agency in the African Informal Economy

What do 'people at grassroots' or 'ordinary citizens' do every day in Africa? The majority spend a large part of their very long working days seeking a living in the informal economy. Statistics tell us that in no other region in the world is the informal economy as central to livelihoods and economic growth as in Africa. Therein the share of employment in the informal economy as a percentage of non-agricultural employment is over 70 per cent, and the share of the gross national product (GNP) produced in the informal economy is over 40 per cent.[5] Central to the argument that there is more to the African city than dystopia is a hopeful reading of the potential of informal economic activities. As Simone puts it, such informal activities 'might act as a platform for the creation of a very different kind of sustainable urban configuration from those currently generally seen' (Simone 2005: 4).

It is therefore important to explore how such an approach conceptualizes economic informality and on the basis of what evidence. Take, for instance, Simone's writing on 'The Informal' in Pikine, Senegal. A description emerges from anecdotal accounts which together provide an image of an informal economy rich with examples clearly belonging to Davis's 'Museum of Exploitation' (Davis 2006: 186). One reads (Simone 2004b: 31, 38, 42, 62) the story of a driver of a local bus accidentally injuring a child while 'veering off into a street-side market stall'. He was killed in retaliation and the vehicle was set on fire. One learns that 'in some areas of the city, up to 40 percent of children do not attend public schools because they are unable to pay even minimal

---

[5] Recent research has importantly questioned the reliability of several datasets on African development (Jerven 2013; Jerven and Johnston 2015). While one needs to be wary of taking any statistical information at face value, there is no denying that the size of the informal economy is very large in Africa.

fees'. Through child labour, these children earn their education in 'informal schools', which Simone suggests to be, perhaps too benignly, 'an important context of instruction'. The reality of widespread under- and unemployment is visible, as 'most young people are, in the end, not working, although they have small opportunities now and then', and so is people's pessimism about the possibility for change, as 'in Pikine everyone, from mayors to grassroots activists, complains about how difficult it is to change anything'. None of the vignettes presented supports the overall conclusion to the chapter, and its paradoxically upbeat tone, that 'in the interstices of complex urban politics, new trajectories of urban mobility and mobilization are taking place', and there are no clues as to how 'a very different kind of sustainable urban configuration' might flourish out of the informal economy. Conceptually, everything seems to fall into the murky category of 'sustainable urban configuration', including urban contexts in which poverty sustainably (or chronically?) reproduces itself. In sum, the very incapacity to take structural forces seriously that characterizes Simone's contribution to the study of the African city reveals itself through, and fuels, his analysis of the informal economy. This amounts to a romantic celebration of survivalism and of 'people's agency'.

In a similar vein, Pieterse (2008: 114–15) suggests that:

Clearly, for as long as the contemporary capitalist system persists, . . . uneven and highly *exclusionary* (emphasis added), it is likely that it will serve the interest of the majority of the poor better to retain their *autonomy* (emphasis added) and cannibalize formal systems and resources where they can; and to bend parts of the city to their will by defending and consolidating their micro-gains they succeed by simply surviving the viciousness of the city's economic reproduction.

Two aspects are noticeable about the above claim that autonomy from the formal sector and exclusion from the capitalist system are the drivers of the poor's reliance on informal economic activities.[6] First is the lack of evidence to justify it. Second is that it is a way of conceptualizing economic informality that echoes that put forward by the single most influential writer on economic informality, Hernando de Soto (1989, 2001), to whom the analysis now turns.[7]

---

[6] Another important strand of work suggesting, perhaps too hastily, that the problem poor people face is not one of adverse incorporation and exploitation by capitalism through informal and highly precarious work, but rather one of exclusion from capitalism, is that of advocates of social protection, such as Ferguson and Li. They argue that, increasingly, the suffering of the poor has to do with their 'limited relevance to capital at any scale' (Li 2010: 67) and with their 'slim prospects of ever entering the labour market at all' (Ferguson 2015: 12).

[7] My point here is to draw attention to the striking similarities in these authors' conceptualizations of economic informality as a space of autonomy for poor people and not to equate the thrust of their overall contributions. While de Soto's work is to be understood as market triumphalism and fundamentalism, work by Pieterse and Simone cannot. Theirs is a more ambivalent position, in which a critical stance towards the impact of economic liberalization on

De Soto has been the 'self-appointed spokesman of the informal sector' (Breman 2013b: 254) for over twenty-five years, first through his work on Peru (de Soto 1989) and, more recently, through his work on 'why capitalism triumphs in the West and has failed everywhere else' (de Soto 2001). His approach, labelled by many as dualistic, puts forward a simple idea, and a very seductive one for policymakers, that the informal sector exists in *separation* from the formal sector, and it is the refuge of the urban poor who are *excluded* from the formal economy. Reference to the informal poor's agency characterizes this approach also, although its conceptualization of agency differs. There is a pronounced market fundamentalism at work here, as the capacity of the poor to respond to market opportunities explains their resilience and their capacity to survive exclusion from the formal sector through ingenious and 'heroic' entrepreneurialism. The role played by structural forces in causing economic informality is conspicuously overlooked. De Soto suggests that state regulation of the economy, and the endless red tape with which the poor are confronted to register their small businesses, explain why such a large percentage of economic activity takes place in non-compliance with state regulatory frameworks. As de Soto has trenchantly put it, the poor 'do not so much break the law as the law breaks them' (de Soto 2001: 23).[8] The growth of the informal economy is thus framed as a choice by entrepreneurs, rooted in their spontaneous and collective response to over-regulation by predatory state apparatuses.

Similar to postcolonial narratives on the city, De Soto pays no attention to internal differentiation within the informal economy and to whether such differentiation impacts on how informality works for different kinds of people. This is part of his explicit downplaying of the explanatory power of class. Its usefulness, he suggests, has been made redundant by the (alleged) reality that self-employment and working for a family business are the dominant employment relations in the informal economy. As de Soto put it, 'Marx would probably be shocked to find out how in developing countries much of the teeming mass does not consist of oppressed legal proletarians but of oppressed extralegal small *entrepreneurs*' (de Soto 2001: 229, quoted in Davis 2006: 179; Krueckeberg 2004: 2).

---

life in African cities, and the challenges that this presents for urban planners, sits alongside a mainstream anti-public-sector stance. One example of the latter is Simone's unsubstantiated claim that 'targeted investments in human capital creation, employment and entrepreneurship, largely managed outside the public realm, will result in better health and living conditions' (Simone 2005: 4).

[8] In close alignment with de Soto, Simone states that 'An informal sector is thus partially elaborated because of the excess and inappropriateness of regulations that persist in the absence of systematic and realistic ways of assessing domestic economies' (Simone 2004b: 26).

A wealth of studies on the informal economy in Africa (see, among others, Azarya and Chazan 1987; Tripp 1997; MacGaffey 1991; Hansen and Vaa 2004) draw on and amplify, explicitly or otherwise, de Soto's conceptualization of the growth of informality as a process of poor people's empowerment and his assumption that entrepreneurial self-employment is the predominant employment status in it. The agency of 'informal people' in bringing about deregulation from below is emphasized, as is the importance of a shared moral economy driving the spontaneous, unorganized, but nonetheless effective resistance of the African urban poor to their unresponsive governments. As Tripp has expressed it, in a study on the informal economy in Tanzania which was well received by many, the relaxation of government control over the economy in the 1980s 'indicated the beginnings of some long overdue adjustments in state–society relations that made the state responsive to societal needs' (Tripp 1997: 139).

The influence of these ideas on policymakers cannot be underestimated. Inspired by de Soto's insights, programmes promoting the formalization of property rights and small businesses have been rolled out across developing countries, Africa included. A number of scholars have emphasized the weak evidential basis for the poverty-reducing power of these programmes (Gilbert 2002; Home and Lim 2004; Myers 2008). Their record in unlocking the productive potential of the poor has been far from positive. Tanzania has also launched one such programme, the Property and Business Formalization Programme (MKURABITA, in its Swahili acronym), with the direct involvement of de Soto himself and armies of staff from his Institute for Liberty and Democracy. The land titles of over 219,000 properties were formalized in Dar es Salaam alone over five years. In stark contrast to the scenario envisaged by its proponents, the programme resulted in gentrification as poor residents were forced out of the central areas of Dar es Salaam (Muhajir 2011, quoted in Myers 2011). While microcredit has different roots, it is an intervention that dovetails effectively with the narrative of the informal poor as small-scale entrepreneurs, and on that premise it has attempted to foster their financial *inclusion*, with very mixed results. Above all, this approach, and the interventions that emanate from it, suffers from 'jobs dementia', as Amsden (2010: 60) memorably termed it: the belief that by supplying credit or training to self-employed entrepreneurs, demand for such businesses will, somehow, materialize.

There exists, however, a much less positive reading of the widespread informalization of economies and of the nature of the state–society shifts that it has brought about. In contrast with de Soto and his followers, it emphasizes class and the role of structural forces in causing informalization. Work under the informalization approach, first put forward by Castells and Portes (Portes et al. 1989) with reference to Latin America, emphasized the systematic link between the formal and the informal sector. It stressed that behind the pervasive informalization of economies lay capitalists' new strategies to deal with the crisis in profitability which followed the oil crisis of the early 1970s. The growth of the informal economy here is better understood as a process which reflects the strategy by formal businesses of going informal to avoid the costs of rights hard-won by organized labour (Portes et al. 1989). In the words of Breman, whose fieldwork-based research on the Indian informal economy spans over forty years, the informal economy becomes 'a regime to cheapen the cost of labour in order to raise the profit of capital' (Breman 2013b: 1). The role of the state in explaining economic informalization is also important, although in a radically different connotation to that outlined by the dualistic approach. State complicity with the interests of capital and employers, rather than its over-regulation of the economy, is a necessary condition for the informalization of the economy. These are important points, although what matters is to avoid, once more, a functionalist and teleological reading of the growth of the informal economy entirely as a function of capital's interest. Not every occupation or development in the informal economy can be explained as such.

Crucially, attention to dynamics of socio-economic stratification within the informal economy leads scholars such as Portes and co-workers and Breman to reject the characterization of the informal economy as inhabited by own-account entrepreneurs. This, they state, is misleading, for two main reasons. First, it conceals the significant numbers of workers in the informal economy who are working for other people. De Soto's claim, and the uncritical adherence to it in much research, that 'most of the poor already possess the assets they need to make a success of capitalism' (de Soto 2001: 6), sits at odds with evidence from a number of studies showing that it is precisely the lack of any asset other than their own labour that forces the very poor to sell their labour power, in precarious and vulnerable conditions, to asset owners. This is not to deny that self-employment is widespread in the informal economy, although attention needs to be paid to forms of self-employment which, upon closer scrutiny, reveal themselves as wage labour in disguise (Breman 1996: 8). Rather, the point is to question the rosy and idealistic portrayal of self-employed entrepreneurialism and the lack of attention, once more, to the structural constraints that militate against genuine entrepreneurialism. In light of these, the reality of operating at a very small scale is better understood as survivalism, as saturated markets with low financial entry costs negatively affect (already low) profit margins.

In sum, as Meagher usefully puts it, 'We need to break the black box of the informal economy', 'as the dualistic informality paradigm has become too blunt an instrument to capture the growing complexity and heterogeneity of contemporary unofficial economies' (Meagher 2010a: 16). Rather than assuming what the dominant employment relations are in the informal economy, and its potential for growth, we need to explore 'the organizational dynamics and power relations' within it.

1.4 Class Matters

We return to where our journey started, with my early impression of Dar es Salaam's transport system as 'functional chaos'. A brief overview of key contributions to the study of the African city and its informal economies has revealed how functionality and dystopia—and the various conceptualizations of agency (or lack of them) that they offer—are also central to this literature. A critical review of such contributions has highlighted their useful insights and pitfalls. This book is entitled 'taken for a ride' because its analysis challenges and aims to debunk two misleading narratives that have become dominant in writing on African cities and on their informal economies. The first, which can be dubbed as 'postcolonial', is that associated with the influential work of Pieterse, Robinson, and Simone. Its celebration of the functionality of African cities, and of the agency of the African urban poor, including their initiatives in the informal economy, pays inadequate attention to the magnitude and nature of the structural forces at play in African cities and in the informal economy. It also fails to articulate whether and how the alternative order and social fabric woven by urban poor initiatives help to address the structural problems of African cities or the structural disadvantage faced by the poor. A second narrative is that informed by De Soto's market fundamentalism. Its celebratory reading of informal economic activities as poor people's empowerment is equally unconvincing, as it lacks attention to the structural barriers that the poor are to overcome for genuine empowerment, and as it presents a rosy conceptualization of the nature of markets and how they work. A close examination of the reality of working in the informal economy in public transport in Dar es Salaam, and of the operations of private operators in unregulated markets, will suggest that the superior efficiency of the private individual in 'free markets', which has informed policymaking since the early 1980s, is no more than an ideological article of faith. It will also show that readers on urban Africa and their informal economies are not the only category who have been taken for a ride; passengers' experience of Dar es Salaam public transport following economic deregulation was very different from what the advocates of reforms had promised, as speeding and overloading became the trademark characteristics of the inefficient private operators which supplied public transport to the city.

In this book, the question of the extent to which the informal poor exert any agency in the process of development takes centre stage. Answering such a question in a meaningful way requires greater attention to the insight that 'men make their own history, but they do not make it as they please; they do not make it under self-selected circumstances, but under circumstances existing already, given and transmitted from the past', as Marx famously put it (Marx 1852/1974). Thus, first and foremost, the urban poor must be located within an economic and social context. Making sense of the formidable structural forces the informal poor are up against is crucial to understanding their agency in the process of development, in a less romantic and more grounded fashion.[9] At the same time, as the founders of Marxism reminds us, attention to economic structures needs to avoid determinism and teleology, thus preventing due focus on what agency the informal poor might possess and on the possibilities for change: 'History does nothing, it "possesses no immense wealth", it "wages no battles". It is man, real, living man who does all that, who possesses and fights; "history" is not, as it were, a person apart, using man as a means to achieve its own aims; history is nothing but the activity of man pursuing his aims' (Marx and Engels 1844/2012). It is therefore useful to clarify in what way, in this case study on urban public transport in Dar es Salaam and its development over time, I will endeavour to avoid the teleology and determinism that is suggested, too hastily, to be the defining feature of Marxist political economy, starting with my use of class as a category of analysis.

First, I use class as a relational concept. Ownership of capital (such as land and other assets, most notably buses in this case), as opposed to the sale of labour power to those with such capital, determines people's class position in the structure of a given society, and their economic stakes. The social relations between capital and labour, and the dynamics of class formation over time, are the key drivers (albeit not the only ones) of processes of socio-economic change. Their study must begin with questions such as 'who owns what' and 'who does what' with it (Bernstein 2010: 22–23), as these are key to understanding the nature and actors of capitalist development. Only by posing such questions can one see that the private operators of bus public transport in Dar es Salaam are differentiated between those who own the vehicles and those who work on them. They also reveal that the employment relationship linking these two classes is central to the way in which the service of public transport is provided, with all its tensions. The consequences of unregulated employment relations affect not only workers but also Dar es Salaam's public; speeding, overloading buses, and refusing transport to school pupils are to be understood as practices through which workers respond to exploitation by bus owners.

---

[9] A handful of recent studies document wage work in informal labour markets as a key source of income for the 'urban poor' (see Jason 2008; Lourenco-Lindell 2002; Mitulla and Wachira 2003; Boampong 2010).

Second, paying close attention to the specificity of this case confirms the extraordinary heterogeneity that characterizes the conditions of labour today (Radice 2014). The book draws on Bernstein's conceptualization of this phenomenon in terms of 'classes of labour', whose common ground, notwithstanding their heterogeneity, is the need to secure 'reproduction though insecure and oppressive—and typically increasingly scarce—wage employment and/or a range of likewise precarious small-scale and insecure "informal sector" ("survival") activities . . . in effect, various and complex *combinations* of employment and self-employment' (Bernstein 2007: 6–7). This complexity certainly makes categories such as 'worker', 'peasant', 'employed', and 'self-employed' fluid. At the same time, it should not distract from the fact that sale of labour power is the main source of income to the poorest.

Third, the identity, political consciousness, and capacity to act collectively by 'classes of labour' cannot be 'read off' from their socio-economic position in society in a teleological or functionalist fashion; there are a series of complexities and determinations in moving from the 'social facts' to the 'political facts', as Mamdani (1996: 219) puts it (see also Cooper 1983). Thus the study of labour identity(ies) and the extent to which it is experienced and expressed—or not—in class terms is an important line of enquiry in itself. Along these lines, Harriss-White and Gooptu (2000: 89) suggest that 'class struggle is first a struggle over class and only second a struggle between classes'. They show how and why the informal workforce in India is divided and remains engaged in the 'struggle over class' whilst capitalist classes actively wage an offensive against labour. In contrast to the blindness of much scholarship to class in the informal economy in Africa, their work, as well as that by Breman (1996), Harriss-White (2003), and Lerche (2012), sets the bar very high in documenting the abysmal conditions in which unskilled labour fares in India. It examines the contingent ways in which both labour oversupply and labour fragmentation prevent the experiences of class dynamics from translating into widespread political radicalism. By drawing comparatively on these examples of undogmatic Marxist work on informal labour, the book's main contribution is the investigation of how 'classes of labour' fare in the context of bus public transport in Dar es Salaam.

1.5 Neoliberalism, Post-Socialism, and Public Transport

Neoliberalism is a key concept used in this book to understand Dar es Salaam and Tanzania, its policies and socio-economic conditions, and the changing organization and performance of the public transport system in the city from the early 1980s to 2015. This is an analytical choice that needs qualifying, as there is an intense debate over the usefulness of the term 'neoliberalism' itself.

So what does the term mean and what is at stake in its use? To be sure, the plurality of ways in which neoliberalism is understood makes the concept 'chaotic' (Jessop 2013: 65), 'ill-defined' (Mudge 2008: 703), 'promiscuously pervasive, yet inconsistently defined' (Brenner et al. 2010: 184), and even 'confused and confusing' (Turner 2008: 2). A useful starting point to navigate this confusion is to note that, like history for Marx and Engels, 'neoliberalism per se does nothing . . . Rather *people* act to promote politically and ideologically infused ends, and their actions and the premises that they are based in can be interpreted as part of a broader pattern of social action which one can label as neoliberalism' (Harrison 2010: 28). Neoliberalism is therefore a political project, associated with a set of economic policies, and promoted through agents. Whilst wide-ranging in focus, neoliberal policies are rooted in the idea that the promotion of an individual's economic freedom and private capital is the best way to organize economies and societies. But there is also an ideological element to neoliberalism, a way of seeing the world, developmental challenges, economies, and societies, that purposefully attempts to forge a hegemonic consensus on how the world works, who inhabits it, and consequently on what is to be done to solve its problems (Clarke 2005).

The concept 'neoliberalism' emerged in the late 1970s, to capture the process of deregulation that underpinned the political promotion of the freedom of the individual in 'free markets', the superior efficiency of which was influentially argued by economists such as Milton Friedman and Friedrich Hayek. The first experiments with neoliberalism were in the United States and the United Kingdom where, under the governments of Reagan and Thatcher, radical reforms—centred on extensive economic liberalization and attacks on the welfare state and on organized labour—reorganized the economy following the economic crisis of the late 1970s. At the same time, the purchase of these ideas on international financial institutions gave a foothold from which to attempt to give global reach to the neoliberal project.

Crucially, 'neo-liberalism has never been short of state intervention. Indeed it has positively deployed it to promote not so much the amorphous market as the interests of private capital' (Van Waeyenberge et al. 2011: 8). To further complicate matters, the way in which the interests of private capital have been promoted has changed over the years, leading, schematically, to two distinct phases. The first, dubbed 'shock therapy', or 'roll-back' neoliberalism (Brenner and Theodore 2002), entailed promoting private capital through changes in economic policy that 'rolled back' efforts to manage the market by the state; privatization and economic deregulation were its key manifestations. The second, dubbed as 'roll-out' neoliberalism, saw a different role for the state, which was brought back into the scene to correct market imperfections and to improve its workings, often in response to problems created during the 'shock therapy'. To capture this change with an example, and one that is relevant to this book, whereas in the 1980s neoliberalism in Dar es Salaam public transport manifested itself as budget cuts to the public provider of transport, its retreat from service provision, the opening of this service to private operators, and the progressive economic deregulation of their activity, in more recent times the focus has been on boosting the role of the state to reorganize the provision of this service under a public–private partnership. Boosted by a multimillion dollar loan from the World Bank, the Tanzanian state has embarked upon a major project to radically alter the provision of public transport in Dar es Salaam. The public-sector role in the partnership is to finance infrastructural work and to regulate the activities of transport providers which, as part of the conditions attached to World Bank lending, will be private. Thus, the PPP entails the use of public debt to fund the opening up of public transport in the city to private (and perhaps controversially foreign) capital of various forms.

If the focus of neoliberalism has thus changed over the years, the way in which the project has taken root has been equally uneven. This is because the promotion of this political project, and more specifically the agendas of its agents in individual places/sectors, does not take place in a vacuum. It entails encountering possible or actual resistance to it, leading to context-specific forms of 'actually existing neoliberalism' (Brenner and Theodore 2002). It is this changing focus of neoliberalism over time, its many dimensions and the uneven way in which they have manifested themselves in different places at different times, that has led some to call for an abandonment of the concept, for, the argument goes, it conceals more than it reveals.

Debates on how to study urbanization and urban experience are a useful entry point to further explore this issue. For some, cities are 'major sites of civic initiative as well as of accumulating economic and social tensions associated with neoliberal projects' (Jessop 2002: 455, 452), and places in which their tensions are visible 'more saliently in everyday life'. However, sceptics question 'the extent to which the diversity of urban outcomes with which neoliberal policy circuits are associated necessarily contributes to a wider project of neoliberalization initially specified in its northern idiom' (Robinson and Parnell 2011: 528–9). This is because 'the multiple circulating policy discourses which accompany the international development project, create messy accounts of urban change', processes that 'are not easily attributable to something called neoliberalism'. Other scholars, while equally aware of the murkiness of the concept, have worked with it, for its rejection entails taking 'analytical attention away from the underlying *commonalities*—for instance, an emphasis on market logics, private property rights, economic optimization and commodification—that pervasively recur amid otherwise diverse forms of regulatory experimentation and institutional reform' and 'an embrace of unprincipled variety and unstructured contingency' (Brenner et al. 2010: 203).

The disagreement amongst scholars on the usefulness of neoliberalism as a concept thus centres on questions that have to be validated, above all, empirically. In doing so, the task at hand is, in Hart's words, 'coming to grips with how identifiably neoliberal projects and practices operate on terrains that always exceed them' (Hart 2008: 678). Those who think neoliberalism is a useful concept believe it can explain 'how local transformations relate to broader terms' (Harvey 2005: 87). Those who doubt its usefulness do so on the grounds that promotion of and resistance to neoliberalism are not the main game in town, and an excessive focus on them risks missing out on broader processes of economic, political, and cultural development. Instead, they suggest, our attention should shift towards 'the value of a progressive politics beyond a focus on neoliberalism and anti-neoliberalism' (Robinson and Parnell 2011: 529). The debate therefore boils down to assessing to what extent reality on the ground is illuminated or distorted by the deployment of the concept of neoliberalism and to what extent one can find evidence of 'multiple circulating policy discourses', of which a neoliberal discourse is just one.

This book makes the claim that, from the early 1980s to the present, Tanzania's wider political economy, including changing policies in the public transport system, can be understood as embodying certain logics of neoliberalism. As I will go on to show, there is much to be lost in rejecting the concept of neoliberalism when making sense of how the public transport system in Dar es Salaam functions, or does not function, and the way in which it has changed. The promotion of neoliberalism, in complex and different ways over time, has proved to be the main force behind these changes. It has set 'the political and economic agenda, limit[ed] the possible outcomes, biase[d] expectations, and impose[d] urgent tasks on those challenging its assumptions, methods and consequences' (Saad-Filho and Johnston 2005: 3–4). Indeed, Dar es Salaam's experience of a shift in the provision of public transport from public to private mirrors that of every main African city. However, the actual shape neoliberalism took in Dar es Salaam public transport and its politics was unique. Grounding neoliberalism in this particular context therefore necessitated attention to the domestic politics and path dependency of its promotion, and to the challenges and resistance that this triggered. This required, first and foremost, identifying the range of different actors who had stakes in the sector, and on what sorts of power, if any, they drew in their attempts to influence developments in it.

Post-socialism is an insightful analytical lens to understand neoliberalism, its politics in Tanzania, and the ambivalent positions of the Tanzanian government towards economic reform, including the direction of public transport in Dar es Salaam. The Tanzanian government has been seen as a model implementer of economic liberalization in many ways, but far from straightforward in its commitment to it, as values and discourses from the socialist period both called for and informed its attempts to intervene in the economy. Research on Eastern and Central European countries, and on China, has thrown important light on the hybrid and contradictory thrust of policy-making in countries embracing economic liberalization following the demise of socialism (Burawoy and Verdery 1999; Grabher and Stark 1997; Hann 2002; Stenning 2005; Walder 1995). Among Africanists, Pitcher and Askew incisively argue for more historical depth in much research on neoliberalism in the continent and in particular for an understanding of 'whether, or how, a socialist past might shape a postsocialist present' (Pitcher and Askew 2006: 3). Theirs is a call to explore the 'institutional and discursive legacies' left behind by socialism, and the 'idioms and symbolic frameworks, collective strategies and individual practices' deployed by those who 'interpret, reject or respond to the momen-tous political and economic changes that have occurred' since the end of socialism.

In paying attention to Tanzania's socialist past and what is left of it, one must begin by noting that its ruling party, the Tanganyika African National Union (TANU) first and then *Chama cha Mapinduzi* (CCM), has been in power since independence.[10] Since then, TANU/CCM transitioned from being the spearhead of *ujamaa*, as Tanzanian socialism was named, from 1967 to the mid-1980s, to being the implementer of one of the most comprehensive experiences of economic liberalization in sub-Saharan Africa, and one for which Tanzania became one of the donors' darlings. Throughout the process, the uses of the socialist past have been open-ended, as different actors have exploited in different ways 'the fluidity and elasticity of concepts' (Pitcher and Askew 2006: 5) that were central to the building of a socialist nation under *ujamaa*, such as egalitarianism, social justice, and state ownership. These have been reappropriated and/or reworked, both to resist economic and political liberalization and to justify it. Jamie Monson's (2006) insightful study of resistance to the privatization of the Tanzania–Zambia Railway (TAZARA)—the 'freedom railway' linking Zambia to Tanzania built with Chinese money during the *ujamaa* period—illustrates how socialist values and discourses were invoked to oppose economic reforms (see also Ghanadan 2009; Kelsall 2003). The socialist regime called on the principle of international socialist solidarity when it exhorted construction workers to see the railway that they were building as theirs, and their work as a contribution to a national and regional liberation project.[11] Such language was turned on its head by local leaders and

---

[10] CCM was formed in 1977, the year in which the Afro-Shirazi Party and TANU, ruling parties in Zanzibar and Tanzania's mainland, respectively merged.

[11] TAZARA was also known as the 'freedom railway', as by connecting the Zambian copperbelt to the port of Dar es Salaam, it freed the Zambian economy from reliance on the railways of white-ruled Rhodesia on which it had hitherto depended.

communities who resisted the closing down of stations and other measures to cut the railway's costs and increase its revenue, as well as its rumoured privatization, precisely by arguing that these were steps that took the railway away from the people, instead putting it at the service of the interests of the IMF and the World Bank (Monson 2006). Socialist values and concepts were not exclusively a weapon to resist reforms. They were also reworked by those in power to fit the new direction of policymaking following economic liberalization. For example, during *ujamaa*, 'self-reliance' was to be achieved through the leadership of the state, its ownership of national assets, and its public action (Lal 2012: 214).[12] It was by invoking 'self-reliance' (*kujitemea*) that President Nyerere refused structural adjustment loans, and the forsaking of Tanzania's national sovereignty in setting its own economic policy that such loans entailed (Aminzade 2013: 239). However, following economic reforms, 'self-reliance' took a number of new meanings, and it increasingly referred to the neoliberal entrepreneurial *individual*, whom public officials exhorted to rely on his/her own initiative after the rolling back of the state (Askew 2006: 31; Lal 2012: 230). Later, Nyerere, and the legacy of his values as the founding father of the nation, proved equally fungible, as they were invoked in different and often antithetical ways by the ruling party, its opposition, and more broadly by actors and authors critical of the path of economic liberalization undertaken by the CCM (Becker 2013; Brennan 2014; Chachage and Cassam 2010; McDonald and Sahle 2002).

The economic context, characterized by sustained jobless growth and by the increasing prominence of foreign investors, helps to explain the government's ambivalent stance as regards the direction of policy. Since the turn of the century, Tanzania has been among the fastest-growing economies in Africa and in the world, its GDP having grown annually by more than 6 per cent between 2001 and 2013 (UNDP and United Republic of Tanzania 2015: 23). As such, Tanzania is part and parcel of the 'Africa rising' narrative (*The Economist*, 3 December 2011; McKinsey Global Institute 2010). Such rates of GDP growth granted its leadership the praise and generous support of the international aid community, which heralded the country as an example of what economic liberalization can deliver for growth. However, the impact of over a decade of sustained growth on poverty reduction has been disappointing, as the reduction of the percentage of people living in poverty has been remarkably slow, from 35.6 per cent in 2001, to 33.6 per cent in 2007 and then to 28.2 per cent in 2012 (National Bureau of Statistics 2002; UNDP and United Republic of

---

[12] It is important to note that 'self-reliance' had taken different connotations during *ujamaa* itself, as it referred to both a goal and a practice, and targeted difference audiences, from the *individual* to the *nation*, linked together by a layer of institutions (the district, the ward, the village, and so forth), all of which had to be self-sufficient (Lal 2012).

Tanzania 2015: xii). Earlier enthusiasm by donors about the path taken by the economy sobered into a concern about 'Sustaining and *sharing* (emphasis added) economic growth in Tanzania' (Utz 2008).

The reasons for the slow rate of poverty reduction stem from the composition and nature of economic growth in Tanzania and, crucially, the impact of this on employment dynamics. The relative size of agriculture shrank from over 50 per cent of GDP in 1987 to 28 per cent in 2010, while over the same period industry and the service sectors grew, respectively, from 15 per cent to 24 per cent and from 38 per cent to 48 per cent of GDP (Wuyts and Kilama 2014: 21). However, the impacts of these sectoral shifts on the structure of employment were negligible, as the sectors driving economic growth, mining (as the most dynamic subsector within industry) and service, are economic 'activities with high productivity but low employment generation' (UNDP and United Republic of Tanzania 2015: xv). Their growth thus did little to absorb the rapidly growing labour force of Tanzania. Manufacturing (another industry subsector) did not contribute significantly to job creation either, as it grew at a relatively slower pace than mining and was concentrated in the production of goods with limited value added. As a result, despite their growth, industry and service combined could employ less than 20 per cent of the country's workforce. The vast majority of the labour force worked instead in low-productivity and poor-quality jobs either in the informal economy or in agriculture, the latter accounting for nearly four fifths of employment (Wuyts and Kilama 2014: 34). In sum, the fact that growth in productivity was concentrated in sectors with negligible employment growth goes a long way to explaining why Tanzania's economic growth has been jobless and thus did not result in significant poverty reduction.[13]

Another fundamental characteristic of the Tanzanian economy, together with its jobless growth, has been the growing weight and visibility of foreign investors in the economy following economic reforms, and the legitimacy challenges that this posed to the Tanzanian leadership in the context of economic growth and very slow poverty reduction. The relaxation of state control was a key component of the strategy adopted by the government, under pressure from donors, to attract foreign investors from 1985 to 2000. Successful as this strategy was, it also brought in its wake a growing resentment about the inclusiveness of the post-socialist order of things and in

---

[13] Demographic trends are also significant here. The fact that Tanzania has one of the fastest rates of demographic growth in the world, at 2.7 per cent per annum, is an important part of the story, as it implies that its economic growth needed to be shared among a rapidly growing population, to the tune of 1.2 million births each year (UNDP and United Republic of Tanzania 2015: 18–19). However, it would be misleading to identify demographic growth as a crude explanation of why economic growth has not translated into significant poverty reduction. After all, the economy grew significantly faster than its population every single year from 1995 to 2015.

particular a recrudescence of anti-foreign feelings that have had a long history in Tanzania (Aminzade 2013). Conflict over land in the mining sector and in agriculture, including violence surrounding cases of 'land grab' by foreign investors, was heavily covered by the media, and a subject of debate in Parliament. Foreigners were central to the 'contentious politics of neoliberalisation' (Aminzade 2013: 276), as media, activists, and MPs, from both the ruling party and the opposition, questioned the capacity of the government to safeguard the interests of Tanzanians alongside and/or from those of the foreign actors who disproportionately benefited from economic growth. It is in this context, and in light of the challenges of legitimacy that it posed to the government, that a number of high-profile policy reversals observable from 2000 onwards can be understood.[14] The employment crisis amid economic growth in Tanzania, the salience of post-socialist politics, the growing anti-foreign feeling, and how the government ambivalently navigated these challenges are also central to this story of the changing nature of public transport in Dar es Salaam. How and why these factors played a key role will become clearer through an overview of the chapters in the book, to which I now turn.

1.6 Methodology of the Book and its Chapters

The key ingredients of the interdisciplinary research strategy that informs this book are its reliance on a mix of quantitative and qualitative research methods, my own fluency in Swahili, and a prolonged engagement with the field, consisting of over fourteen months of fieldwork, and spanning from 1998 to 2014. The two longest research spells were in 1998 and 2001–2, together accounting for over nine months of fieldwork. I then carried out shorter research spells, each lasting approximately three weeks, in 2009, 2010, 2011, 2013, and 2014. These allowed me not only to follow up leads from earlier research but also to observe important changes in the Dar es Salaam public transport system over time, which further prompted my attention to the twists and turns of its political economy. In light of this, it is useful to present the main research questions, as they developed over time, alongside

---

[14] Not everyone would agree with this interpretation. A 2013 *Financial Times* special report on the Tanzanian economy, revealingly entitled 'Statist ways hold back progress', contained the observation by a 'leading businessman' that 'If something becomes successful, the government interferes', laying the blame for such interventions on public-sector corruption. Interestingly, such a position echoed that of the World Bank in earlier years. As the results of structural adjustment programmes began to appear less rosy than the World Bank had earlier expected (World Bank 1989), lack of commitment to reforms by 'political elites, who resisted and distorted SAP for self-interest' and not SAPs themselves (Mkandawire and Soludo 1998: 81; see also Olukoshi 2003) were identified as the cause of the problem. Governance accordingly became a new additional area of work of the World Bank (1989).

the research techniques adopted to answer them. They explain how and why the wide range of empirical data that informs the analysis of this book was assembled or generated.

Newspapers, a remarkably underused source in the study of development, were the starting point of my efforts to understand the context of my research, and they have informed the analysis of every chapter of the book, bar one. As public transport is a central economic sector in the life of any city, and as such very much in the public eye, newspaper coverage of the city's public transport issues provided not only the entry point to sketch out the key trends within the transport system, but also important insights into what were its main tensions, as well as details of the names and institutional affiliations of the key players. This is how many, although not all, of the players involved in the politics and/or policymaking on public transport were first identified, to then further investigate their insights through interviews. Other sources and/or research techniques were more specific to individual chapters, the content and sources of these to which I now turn.

Chapter 2 sets the scene, and sketches the structural forces that influenced the demand for, and supply of, public transport in Dar es Salaam from independence, in 1961, to the present. A complex set of causes, both internal and external, prevented the Tanzanian state from increasing the supply of public transport to match growing demand. These are reviewed. The mismatch between transport demand and supply led to chronic transport shortages and ultimately forced the government to abandon its initial refusal to allow the private sector into service provision. Thus, in 1983, *daladala* entered the market, and by the early 1990s they had become the only supplier of public transport. The chapter explores the link between the momentous shifts in policymaking at international and national level, most notably the rise of neoliberalism and the adoption of structural adjustment policies, and the direction of policymaking on public transport in the city. The main research questions were: (a) how the public- and private-sector roles in the provision of public transport in Dar es Salaam changed over time; and (b) whether the outcome of the deregulation and privatization of public transport and the performance of the private sector justified the belief in the merits of economic liberalization held by its advocates. The chapter reveals how urban passengers of public transport in Dar es Salaam have been taken for a ride literally and also in a more figurative sense. The rationale behind the reforms, which centred on the assumption about the superior efficiency of the private sector if left operating in unfettered markets, proved to be an article of faith. The chapter also highlights the contradictory stance of the government towards economic deregulation, the tensions that it generated, and the winding down of the socialist agenda that it entailed. The initial opening to private operators, in 1983, was seen by the leadership as no more than a temporary and pragmatic digression from its preferred model of state-provided public transport. Further, a number of initiatives on urban public transport adopted by the state since the very late 1990s, when the process of liberalization and deregulation reached its peak, revealed attempts by Tanzanian public authorities to reclaim some policy space and their political context. However, these attempts failed, exposing the state's incapacity to address transport problems and to formulate and enforce regulation of the activities of private operators.

In Chapter 3 the analysis zooms in on the economic organization of the private transport sector with an in-depth focus on its labour relations. Asking 'who does what' and 'who owns what' (Bernstein 2010: 22–23), in the bus public transport sector in Dar es Salaam enabled the debunking of the common assumption that entrepreneurial self-employment is the dominant status of employment in the informal economy. Documenting the significance of class differentiation in the sector, the chapter reveals how insecure and informal employment relations are central to understanding the outcomes of deregulation and privatization and its many inefficiencies. It starts by looking at official statistics on the informal economy in Tanzania, which suggest, like everywhere else in Africa, that wage labour is the exception in the informal economy and self-employment the norm, only to show the pitfalls of such statistical information. These statistics are then compared against results from a questionnaire, completed by over 600 operators, that establish the clear distinction between those who own buses, and those who sell their labour power as transport workers. Drawing on grey literature, newspapers, and my own interviews with bus owners, the chapter discusses the socio-economic profiles of the owners and the profitability and constraints of the business. Utilizing interviews with workers, the chapter explores the uneven balance of power between bus owners and informal unskilled wage workers in the context of unskilled labour oversupply. It also traces the pernicious consequences of the lack of regulation of the employment relationship on the workforce itself and on society. Speeding, lethal accidents involving private buses, the overloading of vehicles, and refusal of transport to passengers entitled to social fares, most notably school pupils, are the most tangible ways in which workers exert their agency as a group on the transport system. This is not, however, the dominant narrative as to why public transport is in such a poor state. Over the years, newspaper coverage of transport problems reveals a narrative that criminalizes the workforce, blaming their hooliganism and greed for the inefficiencies of the transport system. 'Life is war', then, as a transport worker aptly wrote on the windscreen of his bus.

Why did workers fail to respond? In Chapters 4, 5, and 6 the focus is on the economic and political sociology of informal transport labour in Dar es Salaam. The benefits of prolonged engagement with the field are visible here, as developments unfolding over the years prompted a rephrasing of the above question over time. Two distinct phases characterized the workers' political stance towards employers and the state. During fieldwork up until 2002, it could be observed that workers fell short of making demands on employers or on the state, although their behaviour did show signs of collective consciousness. Later rounds of fieldwork in the late 2000s, however, found a much-changed political landscape of passenger transport in Dar es Salaam due to the establishment of a transport workers' organization. At least formally, the very presence of an institution, founded by transport workers for transport workers, reversed the political asymmetry that had characterized urban transport policymaking until then. How workers organized then became the main research question. Through fieldwork in 2009, 2011, and 2014, I explored the reasons behind workers' more assertive and successful attitude towards the state and employers, in partnership with the Tanzanian transport union.

Chapter 4 analyses the barriers that prevented between 20,000 and 30,000 informal bus workers from responding collectively to their criminalization and to the economic squeeze by employers. Starting from a reflection on the sources of workers' power (Wright 2000; Silver 2003), and by comparatively drawing on Asian literature on informal labour, the chapter shows how labour oversupply, its fragmentation amongst different 'classes of labour' performing different tasks as 'transport workers', and geographical dispersion explain workers' lack of effective collective response to their plight. The chapter also reflects upon the forms and limits of *existing* workforce solidarity by drawing on the longitudinal study of the rise and fall (1998–2005) of a labour association by transport workers. The study of such an association, consisting of people from various 'classes of labour' in one bus route, aided this line of enquiry in three ways. First, the associational goals that *some* workers set for themselves, and the values that these reflected, make possible an understanding of workers' strategies to negotiate precariousness, and of whether their collective action embodied a moral economy of the poor, and with what potential. Engaging with overly optimistic claims as to the potential for 'everyday' practices by the urban poor to bring about political change, I argue that to stress the capacity of this instance of associationism for 'changing the rules' (Tripp 1997) of the game would be highly misleading. Second, by attending some of its meetings, reading the minutes of others, interviewing its leaders and ordinary members, and observing the daily interaction and tensions between different categories of transport workers across the association's life, the chapter explores the ways in which labour oversupply, labour fragmentation, and the spatial unit of work impacted on workers' identity by eroding their common ground. The study of the association's politics therefore opens a window into the 'struggle over class' (Harriss-White and Gooptu 2000). Last, but not least, the patterns and rhythm of labour circulation *within* the sector are an important element in understanding both the precariousness faced by transport workers and their political behaviour. Access to the association's archive, and most notably its data on the turnover of members, allows some appreciation of the modalities of labour fluidity over time.

Chapter 5 investigates the factors and circumstances that allowed bus workers to switch from *managing* the effects of precarious employment to *challenging* its causes, first through the founding of an association of *daladala* workers, and then through a partnership with the Tanzanian transport trade union. Drawing on correspondence between the transport union and the workers' association, which I was kindly allowed to access, and on interviews with the leaders of the workers' association, leaders of the trade union, and with transport workers themselves, the analysis explores the strategy chosen by workers to make demands for rights at work on employers and the state. The analysis stresses the significance of this case study by engaging with the wider literature on globalization and its impact on labour possibilities, and more specifically on how to organize the unorganized in the informal economy and the goals which workers' political mobilization can (or cannot) achieve in increasingly liberalized and informalized economies.

In Chapter 6 the long-term dynamics of occupational mobility or immobility of *daladala* workers are unpicked. By drawing on the list of the 121 transport workers who were members of the association in 2002, and by tracking their occupational whereabouts in 2009 and again in 2014, the research primarily seeks to answer to what extent work as a *daladalaman*, notwithstanding its hardship and insecurity, fuelled dynamics of micro-accumulation and upward mobility. Semi-structured interviews with twenty-five of these workers aimed to elicit workers' own views on their own occupational trajectory and on the strategies they have used and constraints they have encountered when navigating the labour market.

In Chapter 7 the focus is on the Bus Rapid Transit project (BRT hereafter), the new face of public transport in Dar es Salaam. This was initiated as a public–private partnership, funded by the World Bank. Its implementation dates back to 2002 and it started to operate in 2016. Dar Rapid Transit (DART) aimed to radically transform public transport in the city through large-scale infrastructure work, including the upgrade of over 130 km of the main arterial roads of the city—many of which were to double in width—eighteen new terminals, 228 stations, and the introduction of new buses which would be larger and less polluting than *daladala*, that stood to be phased out of service from the city's main public transport routes. The key research question asks what interests drove the adoption of BRT in Dar es Salaam, and explores the tensions that characterized BRT's implementation and resistance to it. Drawing on newspaper coverage of the project, on literature on BRT across the world, and on interviews with key informants, the chapter discusses the rapid growth of BRTs in the 2000s and 2010s and the vested interests of the BRT evangelical society (composed of international finance and its NGO brokers) which promoted them as the 'win–win' solution to tackle the crisis of public transport in developing countries. Evidence discredits such a 'win–win' narrative by showing what some Tanzanian actors stood to lose from the implementation of BRT in Dar es Salaam and their capacity to resist the project. It is by drawing on a contextualized political economy of the BRT project that one can understand why the project proceeded so slowly between 2002 and 2015. The tensions over the displacement of existing para-transit operators by foreign investors, the inclusion of the existing public transport workforce, employment destruction, and the affordability of the new service, and how these issues were managed by the Tanzanian government, are a window into 'actually existing neoliberalism' and post-socialism in Tanzania.

The conclusion summarizes the key arguments of the book and reflects on the broader significance of this study of power, precarity, and informality in public transport in an African city. Time to depart now, green light, orange light, red light: GO (*daladala* style)!

2

Public Transport in Dar es Salaam

From State Monopoly to Neoliberalism (1970–2015)

2.1 Introduction

This chapter analyses the changing face of public transport in Dar es Salaam from independence (1961) to the present. Dar es Salaam is the largest city in Tanzania, and home to 4.4 million people according to the 2012 Population Census (National Bureau of Statistics 2013). In the early 1980s, mirroring the broader shift from state to market taking place at a national and global level, the government withdrew from direct involvement in the provision of key city services such as waste collection (Kironde 1999; Kassim and Ali 2006), water (Kyessi 2005), and housing (Lugalla 1995). Furthermore, it progressively withdrew from regulating the activities of the private sector in providing these services. Public transport was no exception, as private buses, their number from the early to the late 2000s oscillating between 7000 and 10,000 buses, supplied public transport to the city through 240 routes (see Figure 2.1).

The analysis in this chapter focuses on the range of structural forces that influenced the demand for, and supply of, public transport over time. Attention will also be paid to the rapid growth in private car ownership, as the increased availability of second-hand cars and car consumerism by the newly affluent, a phenomenon aptly referred to as 'cardolatry' in other African cities (dos Santos 2005: 3, quoted in Pitcher and Graham 2006), radically changed the nature of motorized mobility in Dar es Salaam. As I will show, the trajectory of change and policymaking in the micro-cosmos of public transport in Dar es Salaam mirrors the broader picture of Tanzania, and much of Africa as a whole, in its transition from developmentalism, or state-led development, to neoliberalism. The fine-grained analysis of one key urban economic service, in a specific city, allows exploration of the manner in which the state was rolled

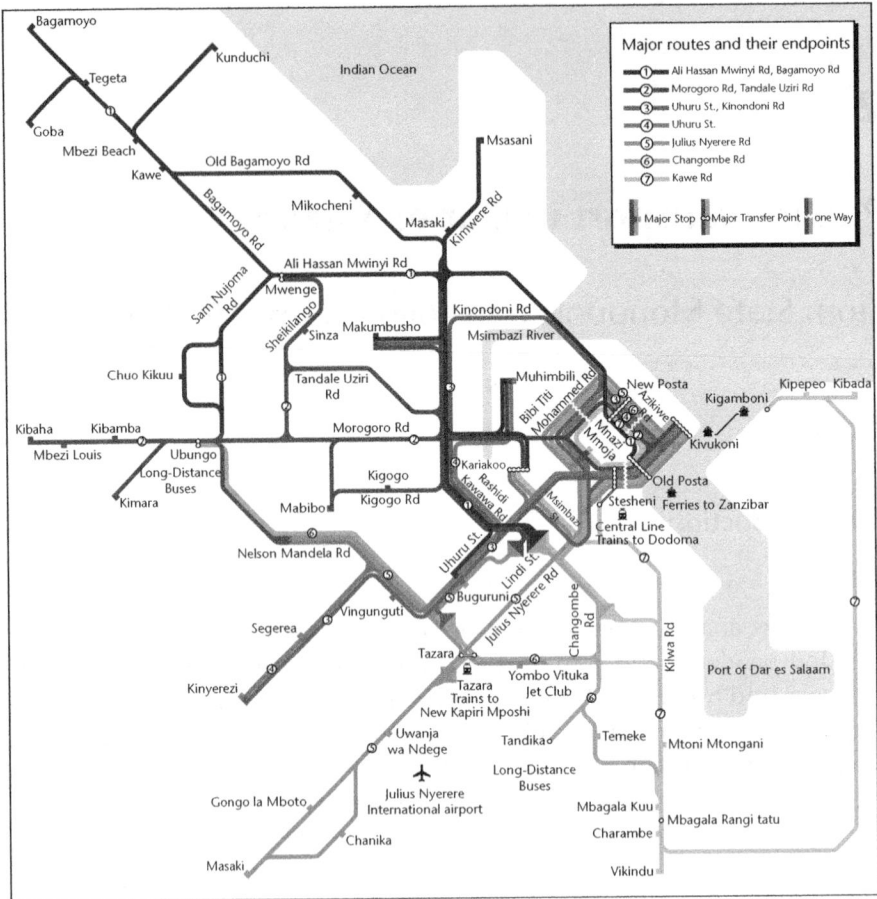

Figure 2.1 Major *daladala* routes in Dar es Salaam, 2011
Source: <ansoncfit.com>

back and of how the legacy of socialism came into play. It also enables us to appreciate the politics, tensions, and path dependency of one instance of 'actually existing neoliberalism'.

2.2 State-Provided Public Transport: 1970–83

With the Arusha Declaration of 1967, the government of Tanzania embarked upon *ujamaa*. Through this, a highly interventionist state placed itself at the centre of a socialist strategy of development. Whole sectors of the economy were nationalized, and in 1970 the Dar es Salaam transport system followed suit with the nationalization of the Dar es Salaam Motor Transport Company

Table 2.1 Public buses in service, 1974–98

Year	Public buses	Year	Public buses
1974	130	1986/7	101
1975	257	1987/8	109
1976	245	1988/9	70
1977/8	221	1989/90	59
1978/9	172	1990/1	32
1979/80	142	1991/2	25
1980/1	141	1992/3	36
1981/2	139	1993/4	54
1982/3	164	1994/5	40
1983/4	141	1995/6	32
1984/5	139	1996/7	24
1985/6	108	1997/8	12

Sources: UDA (1994–8, 1995: 6)

(DMT), a British private company that had operated under a monopoly regime since 1947 (Mamuya 1993: 108). In 1974, the nationalized company was divided into two: the *Kampuni ya Mabasi ya Taifa* (KAMATA) became responsible for interregional routes, while *Shirika la Usafiri Dar es Salaam* (UDA) was granted 'an exclusive licence to operate public omnibuses in Dar es Salaam' (United Republic of Tanzania 1974). The performance of the nationalized public transport company in many respects mirrored the trajectory of developmentalism in Tanzania, which increasingly spun out of control from the mid to the late 1970s. The dramatic decline of UDA is fully documented in Table 2.1, which shows an initial increase in fleet size in the first years after 1974, followed by a steady decrease in the number of buses operated by the company from 1977 onwards. By 1998 only twelve UDA vehicles remained in operation.

Why did the provision of transport by the public sector crumble? Recurrent shortfalls in government funding made available to UDA—which from 1974 to 1983 received only 35 per cent of the foreign exchange funds it had requested from central government (Stren 1989: 52)—indicate the crisis but do not establish its cause. As foreign exchange became scarce, parastatals competing for shrinking government funding responded by pushing up their annual funding estimates, in an attempt to minimize the effect of inevitable government cuts against the budget. Whilst the government's foreign exchange shortage certainly played a role in the UDA crisis, other factors relating to the constraints placed upon the company were of greater significance. Although UDA was formally expected to operate on a commercial basis (United Republic of Tanzania 1974), government in fact privileged the welfare function of public transport by keeping tariffs low and subsidizing the company (Mamuya 1993: 110). As the economic situation worsened, the shortage of foreign exchange made it impossible to purchase the number of buses necessary to match the increasing demand for transport. UDA fleet composition was, in a sense, the legacy of the Cold War. As such, it derived from four different aid programmes funded, respectively, by Hungary, Japan, the United Kingdom, and the Federal Republic of Germany (Wilbur Smith Associates 1991: 3–4) and was further complicated by the presence of five different bus manufacturers (Godard and Turnier 1992: 53). A lack of imported spare parts further took its toll on service provision, delaying the repair of the existing fleet. An additional factor behind the crisis was the poor public management of existing resources. Underqualified personnel and absenteeism were mentioned by UDA as causes of its difficulties in the early 1990s (UDA 1994: 31–2). Furthermore, the average operating life of buses in service was shortened by the deterioration of an inadequately financed urban road system from the 1970s to the 1990s (Kulaba 1989: 240).

While transport supply was in decline, demand was rapidly rising. In 1967, Dar es Salaam had a population of 273,000. According to the National Census, by 1978 this figure had nearly tripled to 769,000 (O'Connor 1988: 136), and in 1988 the Population Census (National Bureau of Statistics 1989) recorded the city as being home to some 1,360,850 residents. Fifteen years later, in 2013, the number of its residents stood at 4.4 million. Government was slow to respond to the increased demand for transport that rapid urbanization implied. Filling this gap, private and informal operators began operating alongside the public company. The government tolerated their activities from 1972 until 1975. However, as soon as UDA was able to double its supply of transport, from 130 buses in 1964 to 257 buses in 1975, such operators were banned (Stren 1989: 52). Opposition to the private sector was at this time deeply embedded in the ideology of the Tanzanian government. In a country whose leadership had set a classless society as the final goal of its development strategy, a highly progressive tax system and frequent increases in minimum wages were policies adopted to reduce the income disparity amongst workers in the formal sector. Tanzania indeed managed to substantially reduce the gap between the maximum and minimum wage: the ratio falling from 1:50 in 1961 to 1:7 in the early 1980s (Maliyamkono and Bagachwa 1990: 30). From this perspective, it is possible to understand why informal economic activities—the profits and wages of which were not subject to the state's redistributing action—were perceived by the authorities as an arena where individuals could profit at the expense of others, as a threat to the establishment of a more egalitarian society, and as a disturbing evidence that the economy was not developing according to *ujamaa* plans. However, the government's adoption of policies that penalized activities within the private sector, maintained until the mid-1980s, ran alongside the decline of state economic capacity. As Mamuya (1993: 111) calculated, in the early 1980s UDA could satisfy only about 60 per cent of the total demand. Thus, authoritarian dismissals of public transport provision by private operators became less and less politically feasible.

Nor was the revamping of UDA through public spending an option. By the early 1980s, the Tanzanian leadership had come to learn how important government fiscal discipline and the rolling back of direct state involvement in the economy, both as a regulator and as economic actor, had become to the World Bank and the International Monetary Fund. The early 1980s were years in which the government attempted to resist cuts in public spending, and to develop homegrown alternatives to structural adjustment policies. It was a short-lived attempt, as Tanzanian leaders were forced to capitulate to demands by the international financial institutions and bilateral donors, and from 1985 structural adjustment proper began (Stein 1991; Wangwe 2004: 394–400). The opening up of the transport sector to private operators took place at this historical moment and its story reflects the leadership's attempt to resist the imposition of economic liberalization, and more broadly the rolling-back of the state from outside, its incapacity to do so, and the partial and half-hearted commitment to reform that derived from these circumstances.

2.3 The Privatization and the Progressive Deregulation of Public Transport: 1983–2001

In 1983, transport service provision was opened up to private operators. This step was taken alongside new initiatives designed to strengthen the public transport supply. The government issued a directive requiring all ministries, government departments, and parastatals to use their staff buses as commercial public passenger buses (Mamuya 1993: 111). As a consequence, sixty-one buses belonging to twenty-seven parastatals began to cooperate with the 141 operational buses of UDA (*Daily News*, 31 March 1983). In the private sector, the government removed import restrictions and delegated the registration of private operators to a committee whose members came from the Ministry of Communication and Transport, UDA, and the National Transport Corporation. Two aspects of the registration procedures have to be stressed in order to understand the political climate and the significance of this opening up to private operators. First, public authorities retained full control of the process by which private operators entered the market. Thus, UDA was given sole authority to legally register all private operators. Furthermore, applications had to be presented to a section in the Ministry of Communication and Transport, and then forwarded to UDA for selection. To obtain a licence, private operators had first to satisfy the safety requirements of the Traffic Police Department. Successful applicants were then allocated a route on which they were allowed to operate as subcontractors, for which they paid UDA a monthly fee. These arrangements were defined in the contract held by UDA (United Republic of Tanzania 1974: 2). Second, the government also retained the power to set fares. Since the UDA licence stated that 'the fares to be charged by the Company shall not exceed those approved by the government or any other authority responsible for setting fare rates', private operators, as subcontractors of the UDA licence, were formally compelled to operate at tariffs set by the authority concerned.

These modalities of regulation reveal that the initial opening up of the Dar es Salaam transport system to private operators took place without a substantial revision of the ideological underpinnings of the monopoly previously held by the state. The public transport company formally maintained its monopoly, while subcontractors were conceded monthly licences, leaving the government with the potential to restore the monopoly at notice of no more than one month. The Tanzanian government therefore, while responding pragmatically to pressure for reform, retained the state's entitlement to regulate entry into the market. It follows that the government at this stage, unrealistically one might argue in retrospect, saw the private sector solely as a means towards a temporary easing of the shortage of transport supply, and not as a long-term departure from the preferred policy of public provision of public transport.

Such was the stance of the Tanzanian government in 1983 and, interestingly, it was a position which survived the country's adoption of structural adjustment policies in 1985. The Ministry of Communication and Transport's proposed national transport policy of 1987, issued four years after the initial opening up of the sector, vividly demonstrates that the preference for public ownership of the service, a residue of the socialist era, still informed the planners' vision of the future in the post-socialist period. First, in clear and apparently unequivocal terms, the Ministry stated that 'whatever the combination of polices adopted in each town, public transport (rather than private transport) will be essential', continuing, 'ideally, public transport, and in this case the bus, is and will always be considered more efficient than private transport' (Ministry of Communication and Transport 1987: 49–50). However, when discussing the specific case of Dar es Salaam, the same document more pragmatically conceded that 'other operators, including private operators, can also be incorporated, but under the coordination of one institution'. Privatization in transport was only to operate under parastatal control, it seemed, and this only within the city of Dar es Salaam.

The state retained not only its entitlement to regulate entry into the market but also the rules of the game within it, including, most importantly, the level of fares. In line with pre-reform principles, the government's fare policy continued to protect low-income groups, over and above fostering private profit. As shown in Table 2.2, although nominal fare levels registered a 600 per cent increase from 1983 to 1991, in the same period the domestic currency

Table 2.2 Bus fares and currency devaluation trends, 1983–91

Year	1983	1984	1985	1986	1987	1988	1989	1990	1991
Tshs/US$	16.34	17.21	38.16	68.65	108.25	154.68	197.00	230.00	300.00
Fare (Tshs)	5	5	5	6	6	8	15	15	30
Fare (US$)	0.31	0.29	0.13	0.09	0.05	0.05	0.08	0.06	0.10

Source: UDA (1994: 22–3)

Table 2.3 Trends in registered private buses, 1983–91

Year	No. of registered *daladala*
1983	178
1984	271
1985	294
1986	300
1987	300
1988	300
1989	175
1990	175
March 1991	355

Source: UDA (1994: 26)

experienced a devaluation of 1836 per cent. As buses and spare parts were necessarily purchased with foreign currency, devaluation severely hampered profitability of the *daladala* business. This goes a long way in explaining why the response of private operators to this new investment opportunity remained modest until 1991.

As Table 2.3 illustrates, official figures show that the number of registered buses fluctuated for the period from 1983 to 1990. In 1983, there were 178 private buses in service. By 1985 this had climbed to 294, only to fall back to 175 in 1989. Interestingly, the first buses in the city to be registered under UDA's new subcontracted licences had previously been operating on upcountry routes. Hence, it was only pre-existing bus operators who were initially able to take advantage of the opening up of the market by shifting their investment to Dar es Salaam. For these operators, Dar es Salaam represented a more lucrative market. For new entrants, however, the policy of protecting passengers via decreases in the real fare level lowered the profitability of the new investment opportunity. Furthermore, existing ownership patterns seem to have had some impact upon trends in registration. According to data from the Ministry of Communication and Transport for 1989, the vast majority of operators owned only one bus. Only two entrepreneurs owned two buses, and only one had more than five (Mamuya 1993: 113). As such, there were very few owners who were capable of increasing their fleet from a business-generated capital base.

The next major twist in the process of reform came in the period 1990–1, when the political mood swung radically towards higher fares and less regulation. It seems that changes in national policy, and a more aggressive phase in the promotion of neoliberalism in Tanzania, influenced public transport policy in Dar es Salaam. In 1989, the final tranche of the 1986–9 structural adjustment programme agreed by Tanzania and the IMF was initially withheld on the grounds that liberalization had to be 'faster and steeper' (Baregu 1993: 111). The World Bank's new emphasis on creating an 'enabling environment' for private investment, which emerged in the late 1980s, rather than the 'shock liberalization' of the early 1980s, had visible repercussions on the direction of policymaking in Tanzania. The government responded to these new pressures with new reforms to the banking sector, to the foreign exchange market, and to the regulation of foreign investment. In the last case, the adoption of the Investment Promotion and Protection Act in 1990 resulted in 'new legal guarantees and certain new benefits' for foreign investors (Gibbon 1995: 13).[1] This meant, amongst other things, full exemption from import taxes, and a five-year exemption from tax for newly established businesses (Baregu 1993: 111).

The Act was accompanied within the Dar es Salaam transport system by the adoption of more profit-oriented fares, and by the abolition of state control on the number of licences issued to private operators. Thus, as shown in Table 2.4 and in Figure 2.2, over the period from 1991 to 1996, whilst *daladala* fares increased fivefold, the domestic currency experienced devaluation of around 100 per cent. The cost of fares in dollars therefore more than doubled over this period.

The powerful effect of this new tariff policy on the registration of private operators was confirmed by the general manager of UDA, Kushoka, who reported that the number of *daladala* trebled in the first six months after the doubling of fares in February 1991 (*Daily News*, 25 November 1991).

Table 2.4 Bus fares and currency devaluation trends, 1991–2009

Date of fare review	3 Feb. 1991	11 Feb. 1993	9 Oct. 1993	7 Jan. 1995	20 Dec. 1996	2007*	2008*	March 2009*	March 2013
Tshs/US$	300	437	530	574	580	1,214	1,140	1,300	1,630
Fare (Tshs)	30	50	70	100	150	250	300	250	400–750
Fare (US$)	0.10	0.11	0.13	0.17	0.26	0.20	0.26	0.19	0.24–0.46

Sources: UDA (1994: 24); *Africa South of Sahara*, 1999; *Daily News*, 8 January 1995 and 31 December 1996; (*) Interview with Sulemani, A., 2010

---

[1] See also Phelps et al. (2007) on the broader practice of investment promotion associated with neoliberalism.

US$

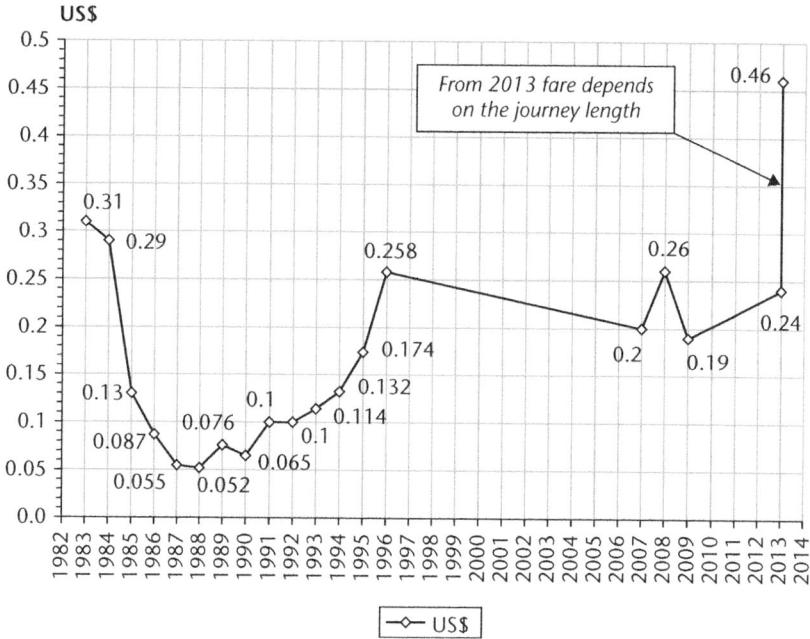

Figure 2.2 Bus fares in US$ in Dar es Salaam, 1983–2013

The shift towards no fare controls in mass public transport in Dar es Salaam occurred in the mid-1990s. More specifically, the transition to a supply-and-demand-driven fare system began in 1996, the year in which the general secretary for the Ministry of Communication and Transport announced that, whilst transport operators could not charge fares above the fixed fare, they could charge less (*Daily News*, 31 December 1996). It was finally completed in 1997 when the minister declared that bus fares would in future be determined solely by market forces (*Daily News*, 25 July 1997).

In 1991 the state also relinquished its control over the licensing of private transport operators, removing UDA's authority to subcontract licences and its power to collect the revenue which it derived from this. The Central Transport Licensing Authority (CTLA) then became the only agency responsible for handling registration applications (UDA 1994: 28), although it had no powers to reject any application except on the basis of non-compliance with roadworthiness rules. These tremendous policy changes account for a dramatic increase in the number of registered *daladala* from 600 in 1991 to 3301 in 1998, as shown in Table 2.5.

By the end of the 1990s, the Dar es Salaam passenger transport system had completed its U-turn from a state-run and regulated market to one almost entirely supplied by private buses with no control over entrants to the market

Table 2.5 Trends in registered *daladala*, 1991–2010

Year	Private buses
(Aug.) 1991	600
1991/2	824
1992/3	1440
1993/4	1484
1994/5	1484
1995/6	1897
1996/7	2342
1997/8	2798
1998/9	3301
2003*	5801
2004*	6600
2005*	7000
2006*	8972
2007*	6144
2008*	5716
2009*	6043
2010*	7573

Sources: UDA (1994: 27); CTLA (1994–8);
(*) Mrema 2011

or over fares. Indeed, its privatization had gone so far that some failed to realize that the state-run UDA was still in operation in the late 1990s. As one of its officials lamented (interview with Mlaki, 1998):

Amongst Traffic Police agents, the younger ones do not know what the UDA is, as nowadays we have very few buses. One day they rang me from the Police Central Station because one of our buses had been stopped by a young agent who believed it was a pirate *daladala* operating without licence. I had to go to the Station to explain to him that we are UDA, that we have a licence allowing us to operate in all areas of Dar es Salaam, and that at one time we were the only supplier on the market.

As Figure 2.3 shows, while the number of officially registered *daladala* continued to rise in the 2000s, there were significant fluctuations in numbers, from a peak of 8972 in 2006 to a low of 5716 in 2008, with an average of 6731 for the period 2003–10 (Mrema 2011: 11).

Meanwhile UDA, the public transport company, was offered for privatization in 1997, although no buyers could be found for over a decade (*The Guardian*, 7 August 2011). The company then continued to supply a negligible number of buses until the early 2010s. Its average bus fleet was thirty-three for the period 2002–8 (National Institute for Transport 2010: 11). Clearly this was insufficient to match the demand for public transport in the city.[2]

[2] As discussed in Chapter 7, UDA steeply increased its fleet and became a much more important player in public transport in Dar es Salaam, following its controversial privatization in 2011 and the beginning of the Dar es Salaam Rapid Transit project (DART).

No. of Buses

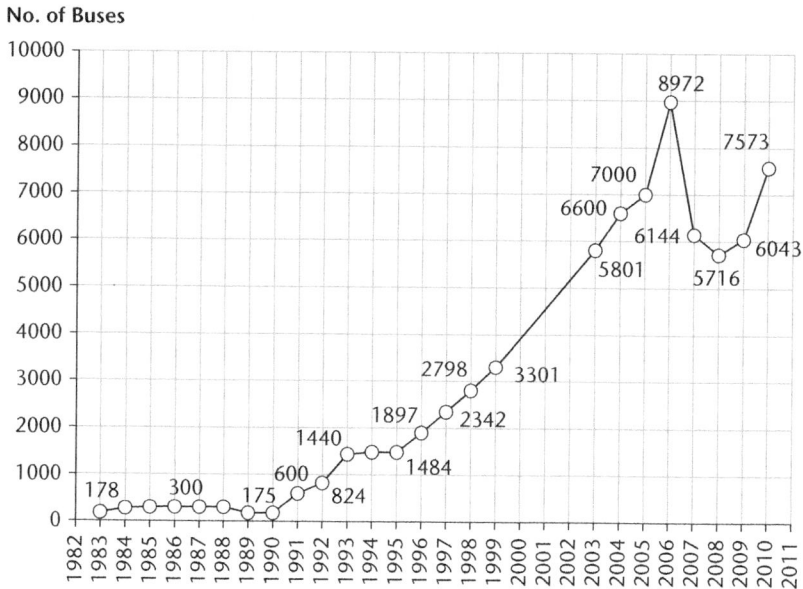

Figure 2.3 Number of *daladala* in Dar es Salaam, 1983–2010

Alongside these trends in the supply of mass public transport by private and public bus operators, the number of privately owned cars rose dramatically from the 1990s and throughout the 2000s. There were about 220,000 vehicles in 2000, and around 650,000 in 2011, with far-reaching consequences for the outlook of transport in Dar es Salaam (Mrema 2011: 10).[3] A number of factors, of a global nature as well as specific to Tanzania, account for this. Globally, since the early 1990s, there has been a marked increase in the availability of cheap second-hand cars from Japan. This has its roots in important changes in the regulation and political economy of Japan's vehicle production. Outstanding progress in the quality and reliability of vehicles produced by Japanese manufacturers presented serious challenges to their profitability—as the life span of vehicles grew and so did the risk of overpro-duction. In response, producers successfully lobbied for new regulations, which made the ownership of older cars in Japan more costly through expensive, and easily failed, inspections (Clerides and Hadjiyannis 2008). Through such efforts, Japanese car manufacturers secured domestic demand for brand-new vehicles in Japan and, consequently, a steady supply of Japan-ese second-hand vehicles entered the international market, a significant por-tion of which were sold in Southern and Eastern Africa (Brooks 2012; Dobler

---

[3] DART (2014) suggests a much lower number of private vehicles, at 120,000 in 2014. No matter which of these figures better reflects reality, the dramatic growth in number of private vehicles and its effects on traffic congestion are not in doubt.

2008).[4] The availability of better communications aided the consolidation of a trade network linking car dealers to buyers in Tanzania. Furthermore, the spread of the Internet allowed prospective Tanzanian car owners to buy directly from Japanese sellers by ordering their cars online or buy locally from a growing army of retailers. Since the late 1990s, prices of cars have dropped considerably due to the influx of Pakistani traders. Charles Mtawali, a Tanzanian car importer and dealer for over twenty-five years, lamented that these traders are first and foremost money launderers, and that they are able to undercut the market price not because they are more efficient traders but because they smuggle drugs to Japan and use the trade in vehicles to Tanzania as a method by which to 'clean' their money (interview with Mtawali, 2014). His claim is difficult to prove. However, it is beyond doubt that the influx of Pakistani traders further increased the availability of cheap second-hand cars in Tanzania, as in other countries in Southern Africa (Brooks 2012; Dobler 2008).

On the demand side, buying your own car in Dar es Salaam became as much a matter of accessing a status symbol as an escape route from the dreadful conditions of public transport provided by the *daladala*. A key factor behind the growth in number of vehicles in the city in the 1990s was the introduction of a monthly car allowance for public-sector workers, twice the value of the monthly wage of the average 'middle-level' employee. The allowance was introduced by the government as a measure to counteract the decline in real wages faced by employees as a consequence of structural adjustment (Briggs and Mwamfupe 2000: 805).

While these factors explain the steady increase in the number of vehicles in Dar es Salaam throughout the late 1990s and 2000s, this growth was not matched by adequate expansion of the road network in the city. Mid-1990s estimates of the rate of growth of traffic density at 6.3 per cent per year were perhaps conservative for the 2000s (JICA 1995, quoted in Kanyama et al. 2004: 13). Traffic in Dar es Salaam has become increasingly congested year on year as the number of vehicles has continued to increase unchecked.

While traffic congestion intensified, new types of public transport emerged, as elsewhere in Africa, such as three-wheeler auto-rickshaws and motorbikes (locally known as *bajaj* and *bodaboda*), which are more adept at weaving through traffic (Anbalagan and Kanagaraj 2014; Goodfellow and Titeca 2012; Chepchieng et al. 2012). The story behind the emergence of the first motorized three-wheeler taxi highlights how individuals responded to the new market opportunities created by the state of public transport provision, and the tensions that this new mode of transport generated.

---

[4] On trade of second-hand vehicles in West Africa, see Beuving (2004, 2006).

The first *bajaj* operator emerged as an unintended consequence of a Tanzanian Red Cross project which aimed to support the livelihoods of a group of disabled people by increasing their mobility through the donation of motorized three-wheelers (Weinstock and Mutta 2009). One of the project beneficiaries, a disabled Tanzanian, Mr Jumbe, initially used the asset to sell fruit at Mwenge, one of the city's largest stations. In the vicinity, there were many destinations which were not served by *daladala* and for which there was a demand for transport. Undercutting taxis, which charged 4,000 shillings (or about US$3) for these short rides, Mr Jumbe began to ferry passengers for about one-eighth of that sum (500 shillings or US$0.40). He was soon imitated by others, and rapidly *bajaj* filled the demand for public transport unserved by *daladala*. Although no authority seems to have information on the exact number of *bajaj*, estimates are in the magnitude of tens of thousands in 2014. In 2010 there were between 200 and 250 *bajaj* imported every month, with between seventy and eighty remaining in Dar es Salaam. Six companies emerged as assembling units of the vehicles, the first one as a subsidiary of a Sri Lankan producer, although the majority were imported from India and China. Over time, as the number of *bajaj* grew, they began to ferry passengers alongside *daladala* routes, and therefore in direct competition with them. Passengers were prepared to pay more than *daladala* fares for their ride, paying a premium for the three-wheelers' capacity to negotiate the traffic more rapidly, often through dangerous driving on hard shoulders, pavements, and unpaved back roads. The same logic applies to *bodaboda* (motorbike taxis), a slightly cheaper and even riskier mode of transport than *bajaj*, for which estimates in 2010 were at 30,000.

As *bajaj* and *bodaboda* established themselves as modes of public transport in the city, the complexity of the sector increased beyond the binary taxi/bus, bringing in its wake a new set of challenges and tensions. First, they both are very risky forms of transport, offering even less protection than *daladala* to passengers in case of accidents. When in a fatal accident a government jeep killed the driver and passengers of a *bajaj*, in 2008, the government banned them as a form of public transport, only to reinstate their licence to operate, following protest, in all areas of the city bar the city centre. But beyond such ad hoc responses to accidents, there was no policy on how different modes of transport were to be integrated in the city. Operators are left to work it out for themselves, often in pretty brutal terms. My direct observation is that the attitude of taxis and *daladala* drivers towards *bajajis* and *bodaboda*, with whom they compete for space on the roads, is extremely aggressive, justifying the existence of 'rumours that some accidents involving *bajajis* have been purposively caused by disgruntled *daladala* and taxi drivers' (Weinstock and Mutta 2009).

Traffic chaos, while partly an outcome of individuals' attempts to escape the mayhem of public transport, further contributed to its worsening quality. Congestion, in particular, spun out of control in the second half of the 2000s. A 2011 study found that, at rush hour, the average speed of a car journey on six routes out of the city centre was as low as 10.1 km per hour (Marcel and Ngewe 2011). Indeed, in my experience, travelling by foot out-performs travelling by vehicle. In 2012, having spent half an hour on a *daladala* to cover the distance between two stops, I decided to get off the bus and walk instead. I was able to cover the 10 kilometres from the city centre to my guest house in about two hours, faster than the bus. Such a state of affairs negatively affected the profitability of the *daladala* business, as operators could only perform a small number of trips each day. Alongside this, currency devaluation increased the cost of purchasing the buses, and of operating them, as the costs of their spare parts and fuel rose over time. In response to increasing costs and decreasing daily returns, the purchase of older and cheaper vehicles has been a widespread strategy of bus owners, as previously discussed. In the second half of the 2000s, two studies indicated that the average age of *daladala* fleets was, worryingly, fifteen years (Dar es Salaam City Council 2006; Kumar and Barrett 2008). Bus workers wrote comments on their vehicles about the state of the buses and their irony is entertaining. One jokingly wrote on the back of his bus *usicheke mgonjwa* ('Don't laugh at the ill person'), while a colleague more proudly asked an imaginary passenger expressing contempt about the state of his bus, *je babako analo?* ('Has your father got one?'). Jokes aside, the ageing of vehicles had serious implications for the deterioration in the quality of public transport. Older and cheaper buses helped to reduce the cost of the initial investment by bus owners, but in the long run this had negative consequences on both the quality and the profitability of bus public transport, as they were more polluting and required more and more resources to maintain.

2.4 Deregulation, Privatization, and the Quality of Bus Public Transport

The analysis has so far highlighted the reform of public transport in Dar es Salaam, the shift from public to private provision of public transport in the city, and the complex factors that explain how public transport changed, alongside the growth of the city and its population, from the 1990s to the present. The chapter now focuses on the effects of the reform on the quality of bus public transport. As the cheapest form of motorized transport available to urban residents, it carries the majority of people.

As we have seen, the reform was premised on the key (neoliberal) assumption about the superior efficiency of the market in both the delivery of services and the allocation of resources within the economy. One way in which extreme competition has been beneficial to passengers relates to fare wars. Private operators were forced to decrease fares in order to remain in competition after the state abolished controls on the tariff. Within one week of government announcements on the liberalization of fare controls, a local newspaper reported that some *daladala* operators had reduced fares from 150 shillings to 100 (*Majira*, 8 January 1997), and I observed during my fieldwork in 1998 that on most routes a fare of 150 shillings was charged only at peak times of the day, while the off-peak fare was settled at 100 shillings. Generally, however, the record of such reforms, as experienced by commuters in Dar es Salaam public transport, paints a less positive picture. Although it is true that the opening of the public transport market to private operators has addressed the chronic problem of an inadequate supply of transport, the lack of controls and regulations on the number of entrants in the market has led to the opposite problem. With the exception of rush hour, bus public transport is very congested, and this results in cut-throat competition amongst bus operators.

The remaining types of competition among private operators were vicious in their effects on passengers, as they entailed a general disregard for the requirements of the safety regulations governing public transport. Overloading of vehicles, divergence from allocated routes, and speeding were practices deemed to be commercially necessary by many operators in order to maintain a competitive edge in a very congested market. In these respects, Dar es Salaam's experience of private provision of public transport mirrored that of other African cities (on Nairobi, see Khayesi 1998; WaMungai and Samper 2006; Salon and Gulyani 2010; on Kumasi, Adarkwa and Tamakloe (2001); on Accra, IBIS Transport Consultants Ltd 2005; on Abidjan, SSATP 2000; Kumar and Barrett 2008; Kouakou and Fanny 2008). The saturation of the market, speeding, and the overloading of vehicles were outcomes of the deregulation of the service.

Unlike other contexts, most notably Kenya and South Africa, public transport in Dar es Salaam was not infiltrated and controlled by organized crime. In Nairobi (Mutongi 2006; Rasmussen 2012) and in South African cities (Bank 1990; Dugard 2001; Khosa 1992, 1994), as well as on transport routes connecting rural areas to cities, research has shown that mafia-style gangs divided up and controlled access to taxi routes, using guns and violence to effectively enforce restrictions on other operators. While no such pattern of serious criminality is evident in Dar es Salaam, 'perfect' competition amongst private operators does have other social costs. As already mentioned, in 1992, 93 per cent of all fatal car accidents in Dar es Salaam involved *daladala* (*Daily News*, 17 May 1993). The use of the word 'accident' in this context is inadequate because it implies misfortune, as opposed to events that are inevitable, and so does not adequately take account of the behaviour of many private operators. The need for excessive speed is also evidenced by the names given to some vehicles by their owners or crew: *Zig-Zag* implying the skill of its driver in overtaking other buses; *Dawa ya moto ni moto* ('If you speed I speed', literally: 'The medicine for the gas is the gas'); *Usipotambaa mswaki* ('If you do not run you will be a brush'), brush being the slang word identifying an empty bus which 'sweeps the road' without any passengers inside; *Mbele kwa mbele* ('Always ahead'), reflecting the daily challenge in the race track that is Dar es Salaam's municipal road system; and finally *Utamaliza kuni kwa kuchemsha mawe* (literally: 'You will burn the whole log to boil stones'), meaning that trying to overtake is as futile as wasting a whole log to cook stones which can never be edible.[5]

To maximize the return on each trip, drivers resorted to speeding. In addition, conductors contributed to maximizing the margin by cramming in as many passengers as possible (*Daily News*, 11 December 1993). This strategy meant that the vehicles would be systematically overloaded. Similarly, attempts to assign *daladala* to specific routes, in order to provide a public service, were ignored by drivers if the returns obtained from operating on quieter, less busy routes were perceived to be too low. With these factors in mind, it becomes clear that many of the inefficiencies characterizing the Dar es Salaam transport system, or the '*daladala* war' as one journalist put it (*Daily News*, 14 September 1991), were a consequence of the competition in the market between private operators. In order to make/maximize profits, private transporters engaged in systematic violations of safety rules and regulations, compliance with which should be the goal of any well-managed passenger transport system.

Another pernicious consequence of cut-throat competition was the daily conflict between *daladala* operators and the beneficiaries of the welfare state. A useful case to illustrate the social conflict generated by the withdrawal of the government from transport provision is the history of the difficult relationship between *daladala* workers and school pupils. The fare policy of the government towards school pupils reflected how the social function of

---

[5] These names, and many others, suggest that a culture of non-compliance is indeed widely pervasive and deeply rooted in city life, a body of rumours, gossip, satire, and new Swahili words witnessing, as Tripp (1997: 179) has put it, 'the gap between party policy and popular sentiments'. However, it is striking that this culture of non-compliance has survived long after the deregulation and privatization of the market, and that such a culture of non-compliance expressed through *daladala* names most often refers to the laws governing speed limits, which cannot reasonably be considered an unnecessary state regulation whose change is long overdue. The necessity of speeding is not shared by all private operators. One *daladala* severely condemns the habit of speeding with the phrase *Huo ni utoto tu* ('That is only infantile'), and another adds *Akikuwa ataacha* ('When he grows up, he will stop it'). Another vehicle named 'Shameless' proclaims its difference from the majority of private operators. The same message is given by the buses: *Mwenda pole* ('It goes slowly') and *Tuliza ball* ('Control the ball', invoking a popular footballing metaphor).

Table 2.6 Trends in student and adult bus fares, 1983–96

Date of review of the tariff	8 July 1983	10 April 1988	6 Sept. 1989	2 Feb. 1991	4 Nov. 1991	17 May 1992	8 Oct. 1993	7 Aug. 1994	7 Jan. 1995	20 Dec. 1996
Adults	5	6	8	15	30	30	50	70	100	150
School pupils	1	1	1	5	5	15	20	20	30	50

Sources: UDA (1994: 24); *Daily News* (8 January 1995; 31 December 1996)

the service was given priority over company profit. From 1983, when private operators entered the transport system, the government attempted to compel private operators to adopt the privileged 'social' fare conceded to school pupils on all *daladala* routes in the city—a policy that persisted throughout the pervasive economic deregulation of bus public transport, as shown in Table 2.6. Over the period from 1983 to 1996, the ratio of pupil to ordinary adult tariffs fluctuated from 1:8 in 1989 to 1:2 in 1992. Thus, even at their lowest level of social protection, school pupils were paying half of the fare charged to adult passengers.

However, the burden of the state's policy toward school pupils fell squarely upon the *daladala* operators. The state has at no time subsidized the economic loss incurred by bus owners for the transportation of these children. The arena of conflict between the two parties is thus clearly defined: school pupils are entitled to concessionary tariffs on any *daladala* journey, but bus operators can only grant those concessions at a personal cost to their revenue from their business. Not surprisingly, clashes between *daladala* workers and school pupils have become a common feature of the Dar es Salaam transport system, with incidents frequently being reported in the press. During the 'Day of the African Child' in 1991, for example, a delegation of children urged President Mwinyi 'to offer them protection against *daladala* bus operators' (*Daily News*, 18 June 1991). As local reports reiterated, it was a conflict with victims on both sides.[6] *Daladala* workers have been taken before the courts and fined (*Daily News*, 25 July 1992; *Uhuru*, 11 June 1994; *Majira*, 24 September 1994), while in 1992 a delegation of workers met with the Minister of Home Affairs to find a solution to the abuses suffered by drivers and conductors at the hands of the school pupils (*Daily News*, 10 March 1992). Government officials threatened to withdraw licences from operators for failure to comply with pupil tariffs, and the actual withdrawal of licences was reported (*Daily News*, 27 July 1991, 6 March 1992, 23 March 1993, 24 March 1995; *Nipashe*, 30 March 1995). Police was called to defend school pupils from harassment by bus crews, when they protested over fares (*Daily News*, 11 August 1992, 14 August 1996).

---

[6] See *Daily News*, 5 May 1993, reporting on a conductor beaten to death by a group of students.

Attempts were made to reconcile this conflict by imposing limits on the number of pupils that may be carried by buses on each journey. Until 1993, government regulations stipulated that small buses should carry a limit of up to five pupils per trip, and larger buses up to ten. But on 26 May 1993, the Minister of Home Affairs announced that this was proving insufficient to meet demand, and that *daladala* would be compelled to transport an unlimited number of school pupils until 7:30 a.m. each morning, so that they could attend school on time (*Uhuru*, 27 May 1993). In response to this government decree, no private buses appeared on the roads of Dar es Salaam the following day until after 7:30 a.m. (*Daily News*, 28 May 1993), vividly demonstrating the incapacity of the state to enforce its own regulatory devices.[7] It is also apparent that the state itself was divided over the direction that public policy should take in this area, with different ministries holding conflicting views. In November 1998, for example, a press release from the Ministry of Communication and Transport sought to better protect *daladala* workers and reverse the intentions of the Minister of Home Affairs, by prescribing that 'small buses should carry three students per trip, big buses five' (Ministry of Communication and Transport 1998). In practice, none of these interventions has proved effective, and the tensions between school pupils and bus crews have continued unresolved (SUMATRA 2007; Ntambara 2013). Commenting on this, one *daladala* worker joked, 'they [the government] are strong like *Coca-Cola*' (interview with Airi, 1998): as everyone knows, *Coca-Cola* is a soft drink. Behind his words lies the important point that the lack of state capacity to intervene successfully is key to understanding why public transport reform has been so messy and has fallen a long way short of displaying the superior efficiency expected of the private sector.

The above insight should not be surprising. A consensus has emerged for some time that privatization and liberalization do not necessarily result in the creation of a vibrant private sector characterized by the growth-enhancing competition for which many had hoped (Ariyo and Jerome 1999; Bayliss and Fine 2008; Bennell 1997; Berg 1999; Cramer 2000; Fine 2008; Mkandawire

---

[7] It is worth underlining the changing nature of social conflict in the transport system. In the early eighties, as Tripp (1997: 1) reports, people's notion of justice caused them to spontaneously unite to contest the legitimacy of the banning of private buses:

A group of about forty Dar es Salaam passengers had boarded a privately owned *daladala* minibus to go to work. On their way they were stopped by a police officer. [...] Realising they would be in trouble, the passengers, who up until that moment had been perfect strangers, spontaneously transformed themselves into one big, happy family on its way to a wedding and started singing, clapping, and making shrill, ululating sounds, as is the custom for people on their way to celebrations. The police, unable to charge the driver for operating a bus on a commercial basis, had no choice but to let them go.

Ten years later this notion of justice has been replaced by competing interests embodying different notions of justice, as *daladala* workers' spontaneous refusal to transport pupils shows.

2001). Even the World Bank, a staunch promoter of privatization as a panacea for economic growth in the 1980s and 1990s, was forced to admit that 'competition is more important than privatisation per se' (Campbell White and Bhatia 1998: 24). State regulation, and hence state capacity, might be necessary where social and private costs and benefits do not coincide, to offset the outcomes of perverse competition, and to ensure that a firm acts in a manner consistent with social welfare (Adam et al. 1992: 19). For example, in the case of the privatization of the processing and manufacturing of vegetable oils in Tanzania, Temu and Due (2000: 697) found that whilst short-term profit maximization was achieved through the import of cheaper oils, this hindered the national economy through a lowering of demand for locally grown raw materials, and diminished employment opportunities for local people.

2.5 Feeble Attempts to Regain Public Control: 1999–2015

The impact of far-reaching liberalization of bus public transport can be seen as an example of the way in which neoliberalism is prone to 'spontaneous combustion' (Jessop 2002: 456). The chaotic state of transport in Dar es Salaam—continuously reported by the local press—was/is a daily display of government ineffectiveness and, as such, has prompted public attempts to address it. In this light, one can read a number of the government's initiatives to reclaim policy space over transport matters as signifiers of attempts to move away from the neoliberal direction of policymaking. Regulation, however, requires understanding of market structures, of the reactions of the various actors within the sector, the design of a set of rules, and the capacity to enforce them, seeking to direct actors towards the desired functioning of the market. Effective regulation thus entails both political will and the capacity to enforce the rules that have been set. The analysis now explores the capacity of the Tanzanian state to regulate public transport in Dar es Salaam in light of these two dimensions.

First and foremost, the Tanzanian state displays no capacity to crack down on the existence of significant numbers of *daladala* operators supplying the market as 'pirate' buses, i.e. without a licence. There are no reliable regular statistics on the exact numbers of pirate operators, but two surveys conducted at the beginning of the 1990s suggest that there were then about as many unlicensed operators as those with licences: in 1990 it was estimated that there were 202 unregistered *daladala* operators, and 175 registered; in 1991, the figures were calculated as 338 unregistered and 355 registered operators (UDA 1994: 26–7). An estimate from 1998, made by the president of the *daladala* owners association, suggested that there were as many as 7648 *dala-dala* in operation (interview with Ndaombwa, 1998). In addition, an article published in *The East African* during 1999 provided an estimate of '6,300 plus' buses (*The East African*, 21 July 1999). Both these estimates again suggest that the number of unlicensed buses remains close to the number officially registered, at somewhere between 3000 and 4000. Research in 2003 similarly failed to come up with a more precise figure of operating *daladala* than between 6000 and 7500 (Kombe et al. 2003). More recent estimates of the number of 'pirate' buses are not available, but on every fieldwork spell in 2002, in the late 2000s and early 2010s, I could observe the presence of pirate buses, especially at rush hour in the mornings and afternoons, and late into the evening and night.

It is also important to highlight that the monitoring of supply in the market is far from precise. My own survey of the data of three different institutions concerned with the operation of private buses shows that there is no consistent figure for the number of compliant private operators. In 1998 I collected data on *daladala* registered with the City Commission (to which private operators have to pay a road tax), with the Tanzania Revenue Authority (to which operators pay income tax), and with the Central Transport Licensing Authority (which at the time issued the licences for passenger transport). While the Dar es Salaam City Council (DCC 1998) reported a total of 4012 compliant vehicles, only 3029 private operators are registered as paying income tax,[8] and only 3301 *daladala* are recorded as having been issued a licence by the CTLA (1994–8). These divergent figures suggest that, throughout the 1990s, operators complied with some of the state regulations, whilst ignoring others.[9] The choice between formality and informality depends on entrepreneurs' individual evaluation of the comparative costs of legality and illegality, rather than on their perception of the legitimacy of government regulations. If, for instance, a newspaper publicized that the Traffic Police gave the order to impound vehicles not paying income tax, as reported on *Daily News* (12 March 1996), more operators were likely to opt for complying with this requirement as efforts to crack down on non-payers increased the cost of non-complying. A number of factors explain the failure to eradicate 'pirate' buses from the streets of Dar es Salaam. At rush hour they were needed, so the authorities turned a blind eye. Furthermore, police enforcers on the streets often had a different agenda from those ordering the enforcement of regulations, and thus settled non-compliance by taking cash bribes.

The establishment of a dedicated urban transport institution in Dar es Salaam highlighted the public authorities' intention to respond more forcefully to public transport problems, yet at the same time, the serious capacity constraints that undermined their effort. In July 1999, the CTLA relinquished

---

[8] The Tanzania Revenue Authority does not keep a centralized record of the data. The figure of 3029 is obtained by the sum of the operators registered by Ilala District (1254), Kinondoni District (1058), and Temeke District (717) (TRA 1998a, 1998b, 1998c).

[9] The procedure has changed, and improved, with the establishment of SUMATRA (discussed later in this section) and the centralization of the process of issuing licences.

its licensing mandate in the capital city to the newly established Dar es Salaam Transport Licensing Authority (DRTLA), a move triggered by the Dar es Salaam Regional Commissioner's need for an institution through which to address the problems faced by passengers (interview with Mhina, 2010). This allowed a more proactive approach to the management of transport problems, within the serious limits imposed by staff shortages and a lack of resources. As Tambo Mhina, secretary of DRTLA from 1999 to 2005, recalled:

> We looked at *daladala*, and we asked the CTLA how many there were. They had no statistics. How many per route, they did not know. What is the procedure to give them a licence? We discovered that if a businessman came, bought his vehicle, passed the inspection from the traffic police, paid the Tanzania Revenue Authority, that was it, they would issue a licence to him.

DRTLA leaders began by assessing how many routes were needed and by establishing the beginning and the end of each route. By drawing on existing population statistics, they estimated trends in population movements throughout the day, so that a target number of vehicles operating on each route could be established. Congested routes were declared 'sold out'. Owners who could not be issued a licence to operate on their first-choice route were allocated less well-serviced routes. The rationale was to ease congestion in some areas and avoid shortage in others. These basic measures reflected both DRTLA's attempts at reasserting some degree of public regulation of private transport, and their limited ambition due to lack of resources.

In a similar vein, late in 1999, the DRTLA chairman required buses to have a coloured stripe corresponding to their licensed route (interview with Mwaibula, 2010). While this measure aimed to make pirate *daladala* more easily identifiable, there were other contentious issues—such as the tense relationship between school pupils and bus workers, and the unregulated nature of employment relations in the urban transport sector—which the state, weak due to lack of financial and human resources, left unchallenged. Policymakers also felt that the fragmented ownership of buses was a further hindrance to effective policymaking, as there was no real counterpart with which to engage. As Tambo Mhina recalls, any liaison required a meeting with thousands of separate owners (interview with Mhina, 2010). It proved impossible to convene such a gathering. To move beyond this, DRTLA proposed a tender system for a maximum of two or three organizations to provide transport services in Dar es Salaam. Such a plan did not convince high-level decision makers, who objected: 'These people who own one or two buses, they are original Tanzanians. If you advertise a tender for large buses and large companies, investors of international calibre will come. They [Tanzanians] have already invested in the sector and will be side-lined. Where will they put their capital?'.

A more ambitious step to reclaim policy space on transport matters took place in 2001, the year in which the Surface and Marine Transport Regulatory Authority (SUMATRA) was established by Act of Parliament (United Republic of Tanzania 2001), subsuming DRTLA. The most significant change was that the public sector reasserted its authority to set fares. SUMATRA's goals presented the somewhat unrealistic win–win objective 'to protect both investor and consumer interests' and 'the desire to Promote (sic) competitive rates and attract the market'. In practice, SUMATRA kept fares at Tshs 150 until 2007, then raised them to Tshs 250 partly to reflect the rising operating costs (especially fuel prices) faced by private operators. In 2008 a further rise to Tshs 300 shillings was agreed, taking into account the further rise in oil prices. Reflecting a subsequent drop in oil prices, in March 2009, fares regressed to Tshs 250. However, in 2013 fare levels increased and ranged from Tshs 400 to 750, a historical peak.[10] Thus, despite the government's willingness to reclaim the authority to set fares with SUMATRA, this has not resulted in cheaper fares for passengers (see Table 2.4 and Figure 2.1).

Furthermore, fare levels, as they are nominally set, tell only one part of the story. In non-compliance with such regulatory efforts, direct observation at each fieldwork spell from 2009 to 2014 revealed that many buses, whether operating with a licence or not, demanded higher fares during rush hours.

Such lack of capacity to enforce regulations applies to many other instances of SUMATRA's attempts to improve the modalities and standards of service provision. For example, SUMATRA attempted to decrease pollution and to increase the service life of private buses, by stipulating that no vehicle would be given a licence if it were more than five years old (United Republic of Tanzania 2007: 66). However, SUMATRA discovered owners could not afford to purchase vehicles of this age (interview with Sulemani, 2010). Special financing schemes for prospective buyers were considered, but in the end the age limit was withdrawn. In a similar but less ambitious effort, SUMATRA prescribed that no car could be imported to Tanzania if it was more than fifteen years old. This was equally unsuccessful as owners simply falsified documents indicating vehicle age.

The scope of regulatory efforts seems to be limited to interventions of which 'maladjusted' African states are typically capable (Mkandawire 2001: 306). They require little or no resources, and their enforcement must be easy, both politically and in terms of resources. This is illustrated by the following example. In 2008 SUMATRA announced that small buses, known in Tanzania as *vipanya* (small mice), would not be allowed to operate in the city centre, and that licences would be issued only to transport companies, not individuals. As

---

[10] Since 2013, fare levels have varied according to the length of the route. Residents living further out of the city centre therefore pay higher fares.

Table 2.7 Licensed buses (*daladala*) by size, March 2008–June 2009

Date	Large buses (25+ passengers)	Small buses (up to 24 passengers)
March 2008	1773	3011
August 2008	2002	2978
April 2009	2617	2474
June 2009	2765	2423

Source: SUMATRA (2009)

shown in Table 2.7, progress was made with the phasing out of small buses from the city centre: the easier part of the initiative. However, the push for the establishment of transport companies, clearly a reformulation of DRTLA's idea a decade earlier, and the more ambitious part of the plan, was dropped (*Mwananchi*, 11 July 2008).

In over four decades, from 1983 to 2015, bus public transport in Dar es Salaam experienced a transition from public to private provision and momentous changes in its regulatory regime. Progressive deregulation culminated in the second half of the 1990s, when the public sector relinquished its control over entrants into the market and fares. Such changes mirror the broader transition from developmentalism, or state-led development, to neoliberalism. Records of the privatization and deregulation of public transport in Dar es Salaam suggest that its residents have been, literally and metaphorically, taken for a ride. The supposed superior efficiency of the private sector did not materialize. Cut-throat competition between private operators in a market that became increasingly congested took many forms. This chapter has also analysed the politics of attempts by the public sector to reclaim some policy space against the grain of economic deregulation. These attempts were prompted, first and foremost, by the problems that the privatization and liberalization of public transport in the city created. However, such re-regulatory efforts, far from being able to impact on public transport problems, reveal above all the state's limited capacity to intervene.

3

'Life is War'

Capital and Informal Labour in Bus Public Transport

3.1 Introduction

The previous chapter traced the momentous shift from public to private provision of bus public transport in Dar es Salaam. Transport 'private operators' or 'bus crews' were at the centre of the narrative, as if the implicit assumption and/or claim that informal actors can be usefully understood as a teeming mass of self-employed small-scale entrepreneurs held true. Such a claim is very common in research on Africa, both mainstream and heterodox, as it is pervasively argued that wage employment has become the exception and self-employment the rule, mainly as a result of the growth of the informal economy. For a very recent example, Fox and Pimhidzai, from the World Bank, contrast the situation in OECD countries, where wage employment is the norm, with that in sub-Saharan Africa where, they argue, 'employment takes the form of self and/or household employment, where a task is performed for family profit or gain (including for home food consumption). *Most labour force participants never even enter the labour market* (emphasis added)' (Fox and Pimhidzai 2013: 3; see also World Bank 2012: 5).[1] As I have shown in Chapter 1, theirs is a widely held belief shared by policymakers. On the basis of this belief, programmes which formalize property rights, micro-finance, and small-scale enterprise promotion are justified.[2] A wealth of statistics, suggesting that working in one's own business is the almost exclusive type of

---

[1] Interestingly, important advocates of social protection, such as Ferguson (2015) and Li (2015), similarly downplay the significance of informal wage labour for the poor, whom they see as suffering exclusion from capitalism rather than from disadvantageous incorporation into it. See also Chapter 1, n. 6.

[2] See section 1.3 for a discussion of this.

Figure 3.1 'Life is War'

employment relationship in the informal economy, backs such a policy focus. The most recent labour force survey in Tanzania is no exception to this.[3]

This chapter has two related goals. First, it aims to build on the picture about bus public transport in Dar es Salaam presented in the previous chapter, and to deepen it by 'unpacking' the private sector that dominates the provision of the service. This entails asking who owns what in the bus public transport sector. Answering such a question opens our eyes to the significance

---

[3] Taking into account criticisms of the narrowness of earlier conceptualizations of informality, there has been greater awareness, at least conceptually, that ' "employment in the informal sector" and "informal employment" are concepts which refer to different aspects of the "informalization" of employment and to different targets for policymaking' (ILO 2013: 33). However, this broader understanding of economic informality has not substantially changed the priorities of policymakers nor the picture emerging from official statistics on the informal economy—which continue to suggest that self-employment is the norm in informal work.

of socio-economic differentiation and of class. Furthermore, the informal wage employment relationships that are predominant in the sector will be shown to be central to understanding why *daladala* operate the way they do, with all their accompanying tensions. The second goal is to question, more broadly, the wisdom of official statistics on the informal labour force. Showing that self-employment is the exception rather than the norm in bus public transport in Dar es Salaam calls into question why informal wage employment hardly exists according to official statistics. Put another way, are the type of employment relationships to be found in *daladala* operations the exception, or is there something very wrong with the way statistics on the labour force are generated?

The chapter starts with this second goal. It answers the above question by moving back and forth from the streets of Dar es Salaam, and the real employment relations to be found in its buses, to a desk, from which a review of existing statistics and debates on *daladala* was carried out. The analysis starts by reviewing some key insights obtained from relevant economic theory, but also from literature on the informal economy in Tanzania. The aim is to unpick the origin of conventional notions of 'wage employment' and 'self-employment', their failure to capture the nature and variety of employment relations in the informal economy, and yet their centrality to the design of workforce surveys in developing countries. The 2006 Integrated Labour Force Survey (2006 ILFS[4] hereafter)—the latest available—is then used to show that the informal economy is seen almost exclusively as a site of self-employment. The analysis then interrogates this claim by looking at the particular type of wage employment relationships that are found in the urban bus transport of Dar es Salaam. To do this, the profiles of the bus owners and of workers are presented, as are the labour relations linking them to one another. The categories and terms with which workers describe their employment situation are then contrasted with the categories and terms used to frame the questions in the 2006 ILFS. The analysis scrutinizes how key employment concepts and terms have been translated from English into Swahili, and how the translation biases respondents' answers towards 'self-employment', thus contributing to the invisibility of wage labour in the collection of statistics on employment in the informal economy.

3.2 Informal Economy as Self-Employment?

Whilst seeing the informal economy as a world inhabited by a teaming mass of undifferentiated, self-employed entrepreneurs is a defining feature of

---

[4] NBS (2007), together with NBS (2009 atd), inform the 2006 ILFS.

neoliberal thinkers—à la de Soto (1989)—it is a way of thinking about economic informality that has a longer tradition, and one that has influenced the design of labour force surveys for many years. Indeed, it dates back to the first time in which the concept of the informal sector was coined. Keith Hart, its inventor, claimed that the 'distinction between formal and informal income opportunities is based essentially on that between wage-earning and self-employment' (Hart 1973: 68), a dichotomy that has been relentlessly adhered to by policymakers in developing countries. In clarifying this distinction between wage labour and self-employment, Hart argued that 'the key variable is the degree of rationalization of work—that is to say, whether or not labour is recruited on a permanent and regular basis for fixed rewards'. Hart's restricted definition of 'wage labour' as permanent and as regular recruitment for fixed rewards is plausible when it comes to describing the nature of the employment contracts in the formal sector. This is the conventional or 'formal' definition of wage labour, which generally refers to 'workers on regular wages or salaries in registered firms and with access to the state social security system and its framework of labour law' (Harriss-White and Gooptu 2000: 89). Production based on this type of 'formal' wage labour is only viable, however, under conditions where productivity is reasonably high and stable relative to the fixed wage rate. 'Formal' wage contracting is unlikely to be widespread under conditions where labour productivity is low, volatile, or unpredictable, which are precisely the conditions that prevail so widely within informal economies in developing countries.

Nevertheless, it does not follow from this that all activities within the informal economy are based on self-employment and, hence, that the capital/labour relation ceases to exist or does so only marginally. Interestingly, Hart gave a detailed account of the variety of production forms that exists in the informal economy: 'In practice, informal activities encompass a wide-ranging scale, from marginal operations to large enterprises' (Hart 1973: 68–70). Yet, surprisingly, he did not draw the obvious conclusion that these varied and often highly differentiated forms of production must imply the existence of a variety of labour regimes, including various forms of wage labour. Part of the problem is that Hart explicitly excluded from his analysis 'casual income flows of an occasional nature', yet recognized that 'some may be hired to small enterprises which escape enumeration as establishments'. He nevertheless went on to say that 'the ensuing analysis is restricted to those who, whether working alone or in partnership, are self-employed'.

In making this restriction, however, Hart fell prey to the fallacy of 'misplaced aggregation' (a term borrowed from Myrdal 1968: Appendix 3): that is, conceptually conflating entities that do not belong together and, thus, should not be aggregated into one category. Indeed, the catch-all category of 'self-employed' conveys a connotation of an individual's own business and/or a family business, of asset ownership, however limited, and of entrepreneurship and some degree of economic independence (Harriss-White and Gooptu 2000: 91). Yet, as Breman argues, 'what at first sight seems like self-employment and which also presents itself as such, often conceals sundry forms of wage labour' (Breman 1996: 8).[5]

The assumption that self-employment is the prevailing status of employment in the informal economy is shared by a number of authors, despite taking antithetical positions in the debate on the potential for growth of the informal economy in Tanzania. Maliyamkono and Bagachwa (1990), Sarris and Van den Brink (1993), and Tripp (1997), for example, who viewed the informal economy as a 'desirable sector', and Jamal and Weeks (1993), who held the opposing view of the informal sector as a last resort, did not question whether there was more to economic informality than self-employment.[6] More recently, along the same lines, in their analysis of labour market dynamics using the Tanzanian urban household panel survey, Quinn and Teal (2008: 4) see the dichotomy between formal and informal employment as identical to that between wage earners and self-employment.[7] That labour regimes in informal production vary widely in nature is thus left out of the picture altogether.

In contrast, the analysis of the informal economy in urban Senegal by Lebrun and Gerry (1974) provides some useful handles to tackle this question since they focus on differences in forms of petty production, from artisans through petty commodity production, to small capitalist production, thus drawing attention not only to the level of labour earnings within the informal economy, but also to the variety of forms of employment. Moreover, these forms do not coexist in isolation, but give rise to a variety of forms, including

---

[5] Interestingly, not every mainstream economist would agree with the view that in sub-Saharan Africa 'most labour force participants never enter the labour market' (Fox and Pimhidzai 2013: 3). Fields (2005: 4), in contrast, models the distinction between the formal and the informal 'murky' sectors, as labour market segmentation or fragmentation: a dualism that implies that different workers are paid different wages in different sectors for comparable work. Fields, however, defines a 'job' for which a wage is paid as a convenient shorthand for 'both self-employment and wage employment', and, hence, blurs the distinction between different forms of employment. This simplifying assumption of labour market dualism allowed Fields to explore distinctive analytical models of the dynamics of labour earnings in the informal sector: more specifically, whether the informal economy is a free-entry sector of last resort or whether it is a desirable sector for employment in its own right, or, as Fields contends, some combination of both, implying internal dualism within this sector (Fields 2005: 17–25).

[6] See Wuyts (2001: 424–31) for a discussion of the underlying models of informal-sector behaviour that underscores these two contrasting views.

[7] In their own words: 'the distinction between formal and informal employment is fundamental to understanding the Tanzanian labour market' (Quinn and Teal 2008: 4). In their survey, all income earners 'were required to assign themselves to one of two mutually exclusive categories: wage-earners and the self-employed', from which they concluded that, 'it is clear that informality is a key characteristic of the Tanzanian labour market: approximately two-thirds of interviewed respondents in 2004 reported being self-employed'.

varied forms of wage labour. Within this spectrum of petty production, some forms lean more towards independent production (which can be best characterized as self-employment, possibly involving the employment of wage labour), while others lean more towards labour contracting. In the former case, it is the product that becomes a commodity; in the latter case, it is labour power that is being (sub)contracted. For example, the itinerant street vendor buys commodities in small quantities from a supplier and sells them to the customers on the streets. Notwithstanding the asymmetric relation that often exists between the vendor and the supplier, this can best be characterized as petty commodity commerce through self-employment. A similar situation prevails for the producer-vendor of foodstuffs—for example, selling meals at the roadside or running a catering service, which may also involve the employment of wage labour (which can be characterized as self-employment using paid labour).

However, many other informal economic activities mainly involve the contracting of labour, and not the sale of commodities. In Tanzania, as shown in section 3.3, employment in informal mining is almost exclusively classified as self-employment. But, as Wangwe (1997), Jonssøn and Bryceson (2009), and Jonssøn and Fold (2009) show, the reality on the ground is much more complex. Production relations in informal mining are distinctly hierarchical, involving claim holders (those holding the mining licence), pit owners (those operating the pit, including the recruitment of labour), and various types of workers. Typically, these workers are not paid a fixed wage, but a share in the output produced, net of 'coverage of workers' reproductive costs (food, medicine, basic health services, and pocket money)' (Jonssøn and Fold 2009: 217). A basic wage is paid, therefore, to cover reproductive costs. Of the remainder, the claim holder usually takes 30 per cent of net output, the pit owner 40 per cent, and workers share the remaining 30 per cent.

Similarly, in the construction industry, informal production has taken on an increasingly prominent role both in output and in employment, in part because of 'an increase in sub-contracting by the formal sector and a new role for the informal sector as supplier of labour', but also because 'an increasing number of building clients are choosing to by-pass the formal sector altogether, and engage directly with enterprises and operators in the informal sector' (Wells 2001: 270). To recruit labour, the client pays for the building materials and engages a 'labour contractor' to supply the necessary labour but, at times, the client directly recruits labour, in which case the contractor effectively becomes a foreman.

These examples illustrate that lumping together the varied forms of labour contracting that exist within informal production, into a single category of self-employment, hides more than it reveals. However, as will be argued in section 3.3, this is precisely what labour force surveys tend to do.

3.3 The 2006 Integrated Labour Force Survey: Definitions and Patterns of Employment

Labour force surveys are among the least frequently carried-out surveys in sub-Saharan Africa (SSA). Since the 1980s, international donors have directed their support towards income and expenditure and household surveys (Oya 2013: 257). Against this trend, Tanzanian authorities have done relatively well, as three labour force surveys were completed in 1990/1, 2000/1 and 2006, and a fourth was in preparation in 2014. Although the quantity of available data on labour is higher in Tanzania than elsewhere in SSA, the quality of such data is low, for reasons that will be explained.

The 2006 ILFS allows us to explore different ways of looking at informal employment using different sets of classifications of the structure of employment: in particular, by industry, by sector, and by status. A further distinction is made between main and secondary activities of employment but, for the purpose of aggregation, only main activities are included to avoid double-counting. The definition of informal sector relates to the type of enterprises, while that of self-employment to the status of employment. Following prevailing ILO guidelines, the 2006 ILFS in Tanzania defines the informal sector as 'a subset of household enterprises or unincorporated enterprises owned by households' (NBS 2007: 7–8). These enterprises 'may or may not employ paid labour and the activities may be carried out inside and outside the owners' home'. The informal sector comprises both informal own-account enterprises as well as enterprises of informal employers: the former employ workers on a continuous basis, while the latter employ workers on an occasional basis or make use of the employment of unpaid family helpers. This definition of the informal sector, therefore, does not exclude the employment of wage labour.

Self-employment constitutes one of the four categories of the status of employment, alongside paid employee, family worker, and traditional agricultural worker. More specifically, those in self-employment are defined as 'persons who perform work for profit or family gain in their own non-agricultural enterprise, including small and larger business persons working in their own enterprise' (NBS 2007: 7–8). This category is subdivided into those with employees and those without employees.

A cross-tabulation of employment status against sector of main employment for the 2006 ILFS is presented in Table 3.1. It shows that 'paid employment' accounts for only 0.7 per cent of informal-sector employment (main activity only) and, hence, is deemed to be a very rare type of employment relationship. Self-employed workers without employees constitute the dominant type of employment status, at 83.8 per cent. Together with self-employed workers with employees, at 13.8 per cent, self-employment totals a staggering 97.6 per cent of employment in the informal sector. This also

Table 3.1 Employment status by sector of main employment, 2006 (main activities)

Employment status	Sector of main employment					
	Central/local government	Parastatal	Agriculture	Informal	Other private	Household economic activities	Totals
Paid employee	439,355 *100.0%*	66,307 *100.0%*	0 *0.0%*	12,274 *0.7%*	1,206,395 *84.2%*	31,563 *6.1%*	1,753,481 *10.5%*
Self-employed (non-agric.) with employees	0 *0.0%*	0 *0.0%*	0 *0.0%*	232,334 *13.8%*	66,552 *4.6%*	899 *0.2%*	299,786 *1.8%*
Self-employed (non-agric.) without employees	0 *0.0%*	0 *0.0%*	0 *0.0%*	1,409,698 *83.8%*	99,828 *7.0%*	3025 *0.6%*	1,512,551 *9.1%*
Unpaid family helper (non-agricultural)	0 *0.0%*	0 *0.0%*	0 *0.0%*	29,366 *1.7%*	61,035 *4.3%*	485,974 *93.2%*	575,798 *3.5%*
Unpaid family helper (agricultural)	0 *0.0%*	0 *0.0%*	1,316,724 *10.5%*	0 *0.0%*	0 *0.0%*	0 *0.0%*	1,316,724 *7.9%*
Work on own farm or *shamba*	0 *0.0%*	0 *0.0%*	11,168,792 *89.5%*	0 *0.0%*	0 *0.0%*	0 *0.0%*	11,168,792 *67.2%*
Total	439,355	66,307	12,485,516	1,682,383	1,432,370	521,202	16,627,133

Source: Elaboration by Rizzo et al. (2015) from the 2006 ILFS (Table 5.8; Tables B4 and B5, pp. 38 and 119)

Table 3.2 Structure of employment by sector, male and female, 2006 (selected subsectors, main activity only)

Industry	Currently employed population (main activity only)					
	Total			Informal		
	Male	Female	Total	Male	Female	Total
Agriculture/hunting/ forestry and fishing	5,880,789 *72.7%*	6,832,446 *80.0%*	12,713,234 *76.5%*	13,296 *1.4%*	6202 *0.8%*	19,498 *1.2%*
Mining and quarry	72,862 *0.9%*	11,463 *0.1%*	84,325 *0.5%*	39,987 *4.3%*	7492 *1.0%*	47,478 *2.8%*
Manufacturing	272,872 *3.4%*	161,335 *1.9%*	434,206 *2.6%*	133,470 *14.4%*	109,533 *14.5%*	243,003 *14.4%*
Construction	171,995 *2.1%*	6686 *0.1%*	178,681 *1.1%*	50,699 *5.5%*	412 *0.1%*	51,111 *3.0%*
Wholesale and retail trade	750,999 *9.3%*	518,357 *6.1%*	1,269,356 *7.6%*	538,496 *58.1%*	428,990 *56.8%*	967,487 *57.5%*
Hotels and restaurants	86,882 *1.1%*	240,552 *2.8%*	327,433 *2.0%*	46,746 *5.0%*	170,387 *22.6%*	217,132 *12.9%*
Transport/storage and communication	231,116 *2.9%*	13,111 *0.2%*	244,227 *1.5%*	25,968 *2.8%*	17,081 *2.3%*	43,050 *2.6%*
Other community/social and personal service activ.	79,336 *1.0%*	35,206 *0.4%*	114,543 *0.7%*	78,789 *8.5%*	14,835 *2.0%*	93,624 *5.6%*
Totals	8,086,325	8,540,809	16,627,133	927,452	754,932	1,682,383

Source: Elaboration by Rizzo et al. (2015) from the 2006 ILFS (Figure 5.2, p. 35, Table B3, p. 118, and Table C2, p. 119)

reveals an interesting anomaly in these data: while paid employees constitute only 0.7 per cent of the total, the self-employed with employees account for 13.8 per cent. Assuming that the self-employed with employees employ at least one employee each, these figures appear to hide the importance of paid employment. Crucially then, as far as the statistical evidence is concerned, informal-sector employment essentially equals self-employment, notwithstanding that the definition of informal sector conceptually includes employment statuses other than self-employment.

Table 3.2 gives a more detailed breakdown of employment figures for selected subsectors of employment for the 2006 survey. The selection of sectors was confined to those with significant employment in the informal economy, although the aggregate totals give the total employment across all sectors of the economy. While the table shows that the informal sector is mainly concentrated in trade, followed by manufacturing, it is important to note that Tables 3.1 and 3.2 feature employment totals by main activity only.

The labour force data also gives information, albeit less detailed, on employment in secondary activities for selected subsectors. As shown in Table 3.3, in 2006, 48.6 per cent of employed persons were engaged in secondary activities (NBS 2007: 52).

Table 3.3 Structure of employment by sector, male and female, 2006 (selected sectors, secondary activity only)

Industry	Currently employed population (secondary activity only)					
	Total			Informal		
	Male	Female	Total	Male	Female	Total
Agriculture/hunting/ forestry	1,218,842	573,391	1,792,234	120,175	18,538	138,714
	35.9%	12.3%	22.2%	10.7%	1.8%	6.5%
Mining and quarry	256,669	301,134	557,803	209,572	273,729	483,301
	7.6%	6.4%	6.9%	18.7%	27.2%	22.7%
Manufacturing	1289		1289	1289		1289
	0.0%	0.0%	0.0%	0.1%	0.0%	0.1%
Construction	625,468	496,099	1,121,567	569,892	458,202	1,028,094
	18.4%	10.6%	13.9%	50.8%	45.5%	48.3%
Wholesale and retail trade	76,501	242,783	319,285	69,289	227,784	297,073
	2.3%	5.2%	4.0%	6.2%	22.6%	14.0%
Hotels and restaurants	51,882	3144	55,026	31,011	899	31,910
	1.5%	0.1%	0.7%	2.8%	0.1%	1.5%
Transport/storage and communication	873		873	16,814	12,026	28,840
	0.0%	0.0%	0.0%	1.5%	1.2%	1.4%
Other community/social and personal serv. activ.	854,801	3,013,198	3,867,999	103,022	16,208	119,230
	25.2%	64.4%	47.9%	9.2%	1.6%	5.6%
Totals	3,397,310	4,677,151	8,074,461	1,121,063	1,007,387	2,128,450

Source: Elaboration by Rizzo et al. (2015) from the 2006 ILFS (Table C2, p. 119 and Table D2, p. 12)

In Table 3.3 the dominant sector appears to be 'other community, social and personal activities' (with 47.9 per cent of employment), the definition of which is left vague in the 2006 ILFS. What is most striking in Table 3.3, however, is the share of employment in mining and in construction. As seen earlier, micro-studies show that in both sectors there is heavy reliance on labour contracting rather than direct commodity production by self-employed operators. Why then, are 97.6 per cent of those employed in secondary activities of the informal sector classi-fied as self-employed by the 2006 ILFS (NBS 2007: 46)? Once more, it is important to note that case studies based on empirical evidence expose that this supposedly overwhelming dominance of self-employment suggested by the statistical evi-dence is by no means straightforward. How then, are these number generated?

One possible reason for the invisibility of paid labour in labour force surveys is that its modules are designed with the realities of advanced economies in mind (Standing 2006). That is, the tools used for surveys on employment stem from OECD economic realities and are not fit for purpose in recording infor-mation about employment statuses for all countries. A recent survey experi-ment by the World Bank in Tanzania aimed to test the extent to which labour statistics are affected by the way in which questions are asked. The experiment included a shorter and longer module to determine employment status.

Although its authors claim that there is a 'significant' impact on results obtained from the way questions on employment status are asked (Bardasi et al. 2010: 25), the picture that emerges from both modules suggests that self-employment remains the norm in SSA. The percentage of people in 'paid employment', for example, varies by a maximum of 5.5 per cent, and as little as 0.1 per cent, but never exceeds 20 per cent. Hence, self-employment, with or without employees, and unpaid family work, when combined, still make up the lion's share of employment, at no less than 77 per cent.

Others have argued that the main problem with the OECD origin of labour force surveys has to do with their definition of paid employment more specifically (Cramer et al. 2008; Oya 2013; Sender et al. 2006). As this is rooted in the conventional conceptualizations of formal wage employment that can be observed in these countries, it is particularly inadequate for capturing informal and precarious forms of wage labour in developing countries. There, the dividing line between wage employment and self-employment is often not as clear-cut as theory and the labour force data suggest. This leads to considerable underestimation of the importance of wage labour in informal production.

3.4 From Statistical Fiction to Employment Realities: The Case of the *Daladala*

A close look at the informal employment relations to be found in public transport buses in Dar es Salaam, to which I now turn, helps not only to substantiate this point further but also to grasp the analytical and policy implications of failing to detect informal wage labour. During my early spells of fieldwork, observation suggested that labour relations were central to understanding the performance of the *daladala* sector. I therefore designed a questionnaire aimed at quantifying the prevalence of different types of employment in the sector, the characteristics of the workforce, and the conditions of and returns from work. In November 1998 I administered the questionnaire which was answered by 668 workers (drivers and conductors represented 48 and 52 per cent of respondents, respectively), randomly selected at four different locations in the city (Mwenge and Ubungo stations, Posta Baharini, and Kariakoo). The sample constituted 3.75 per cent of the estimated total workforce (at 17,800 in 1998).[8] The statistical picture that

---

[8] Although we know the approximate number of buses operating within the market, the number of *daladala* employees is more difficult to quantify. Neither the Ministry of Communication and Transport nor the Ministry of Labour holds any statistical information on the *daladala* workforce. However, as we know there to have been between 6300 and 7648 *daladala* operating in 1998, including both licensed and pirate operators, it is possible to take the mean of

emerges from this questionnaire is compared to the findings of a later questionnaire on the *daladala* workforce in 2003, administered by a coalition including the *daladala* workers' association (UWAMADAR 2003). Through observation (at times participant) and semi-structured interviews with workers from different routes, I both deepened and probed the picture that emerged from the questionnaire, namely the predominance of highly vulnerable informal wage labourers, and explored their employment relationship with bus owners. Attention is also paid to the way in which the dividing line between wage and self-employment becomes blurred, and how workers themselves refer to these employment relations in Swahili. The analysis then contrasts workers' wording of, and thinking about, employment with the words and categories used by ILFS to capture such realities.

Results from the two questionnaires answered by these bus workers in the late 1990s and early 2000s (Rizzo 2002; UWAMADAR 2003) found that family or household employment, so central to mainstream conceptualizations of economic informality, are the exception rather than the rule in this sector. Instead, the *daladala* operations are characterized by a clear division between a class of bus owners and a class of transport workers. Over 90 per cent of the *daladala* workforce, whose total number is estimated to be between 20,000 and 30,000, sell their labour power to bus owners. The vast majority of these workers (82.9 per cent) are employed without a contract (*kibarua* in Swahili) (see Appendix A). Similarly, the 2003 questionnaire, answered by a larger sample, consisting of 2000 workers (1000 drivers and 1000 conductors), found that almost the totality of the workforce, 97.7 per cent, was employed without contracts (UWAMADAR 2003: 11).

Together with class, sex is the other social marker that notably regulates access to jobs in this sector: the workforce is almost exclusively male. In 2003 there were only four women drivers known to be operating in the sector, and I have never come across one during my several fieldwork spells. Interestingly, my discussions with workers consistently found that religion and ethnicity, so pervasive in regulating informal economic activity in Africa and elsewhere in developing countries (Harriss-White 2010), do not play a significant role in this sector. Above all, it is the relationship between bus owners and workers that defines how the public transport service is delivered in Dar es Salaam.

these estimates (6974), and calculate the size of the workforce based on the number of persons employed for each bus in operation. The number of workers per bus varies from two (driver and conductor) employed on small buses (vehicles transporting up to twenty passengers), to three (driver and two conductors) employed on larger buses. Based upon the results of a questionnaire (Appendix A) distributed among bus workers, which received 668 responses, it can be estimated that 44 per cent of buses are of the smaller 'two-man' type, while 56 per cent are the larger 'three-man' type. On this basis, the labour force directly employed on a full-time basis on *daladala* could be estimated at about 17,800 workers in 1998. Such an estimate does not include part-time *daladala* workers of various kinds, the role of whom will be discussed in Chapter 4.

Who are the owners then? Who are the workers? What employment relation-ship links the two groups?

3.5 Bus Owners in Dar es Salaam

The study of the socio-economic profile of the bus owners, and the oper-ational realities they face, is complicated by the fact that individuals, rather than companies, provide the bulk of public transport in Dar es Salaam. This makes it difficult to ascertain whether individuals formally registered as the licence holders are the actual owners or just 'name lenders'. The practice, by wealthy and politically well-connected individuals, of registering buses under their relatives' or overseers' names to avoid unwanted publicity on their investments is well-known (interview with Mayao, 2010; Mfinanga and Madinda 2016: 165), and it has been a source of problems for policymakers on urban transport in Dar es Salaam. As Tambo Mhina, former secretary of DRTLA, complained, 'if you want to meet with owners, it is not easy to get them' (interview with Mhina, 2010).[9] Even assuming that the registered names are genuine, the authorities in possession of the roster of owners (DRTLA in the 1990s and SUMATRA since 2001) understandably could not share it with me.

Nonetheless, a number of insights about bus owners emerge from existing grey literature. First, over the years, a number of studies and reports concur in suggesting that the ownership of buses is highly fragmented. In 1989, data from the Ministry of Communication and Transport indicated that only two entrepreneurs owned more than two buses (Mamuya 1993: 113). In 1994, a survey by UDA stated that the average bus fleet comprised one to two buses per entrepreneur (UDA 1994: 48). More recent studies confirm the same picture. In 2009, a study recorded that only thirty-two licensed operators, out of a total of 4950 owners, owned five or more buses (DART 2010: 11). In 2011, Sumatra, the licensing authority, stated that '80% of individual owners had at most 3 buses (SUMATRA 2011, 33)'. Such claims have to be taken with a pinch of salt, as the concentration of bus ownership might be more significant than these studies suggest due to the above-mentioned strategy by some owners of registering their vehicles under the auspices of 'name lenders'. At the same time, observation of how bus public transport operates allows one to determine that no investor (or group of them) owns a fleet that is large enough to allow them to behave as an oligopolistic supplier in the market.

---

[9] To be sure, owners are organized through the Dar es Salaam Commuter Bus Owners Association (DARCOBOA). However, the association has a relatively small membership, between 280 and 300, out of a possible 4950, at the time of fieldwork in 2009 (interview with Gwao, 2009).

As for the socio-economic background of the owners, the lack of an access-ible and reliable roster of owners similarly complicates access to the group and the gathering of reliable information about them. It is above all a case of rumours, hearsay, and unsupported claims which, as such, needs to be han-dled with care. Nonetheless, local newspapers as well as workers suggest that those with access to public-sector employment are highly represented in the category of bus owners (*Daily News*, 27 July 1991). A 2004 study refers to three categories of bus owners in Dar es Salaam: 'i) retired people, ii) people with low incomes, iii) civil servants', but provides no information of the relative share of each of the three groups, nor of the source for this information (Kanyama et al. 2004: 69). One needs to question how people with low incomes could ever manage to save the sum of over US$10,000 which is required to purchase such buses, especially as another study, drawing on interviews with ninety-nine bus owners, noted that availability of savings acts as a significant barrier to the exclusion from investment in the sector. 62 per cent of respondents drew on their own funds to enter the business. Loans have been accessed by only 28 per cent of bus owners, with the remaining sources of funding being grants of an unspecified nature (at 12 per cent) and 'other' sources (SUMATRA 2011: 37).

The profitability of the business is also difficult to establish, although there are studies that present information on this. In reviewing their insights, it is important to be cautious about the nature of the data that informs them, as owners might be interested in underestimating the profitability of the busi-ness when answering questions from researchers working for the government transport regulator, SUMATRA, or researchers in general. Kanyama et al. stated (2004) that the annual net revenue was between 6 and 7 million a year for a bus bought with 18 million TShs. Their assertion that 'it will take about three to four years for an operator to recover the invested capital' (Kanyama et al. 2004: 73) overestimates the actual amortization time as it is based on conser-vative calculations. Their estimate of annual revenue is based on data for 2003 and 2004, for which they break down expenditure and revenue on a monthly basis. However, six months out of the twenty-four months are left blank and are thus not factored into the calculation of the total net revenue from the two years. If the average monthly net revenue from the eighteen months for which data exist is added to the total, the net revenue from the two years rises from 13 million TShs to nearly 15.5 million. In light of this, the amortization time shortens to two years and four months. A more recent study, from 2011, confirms that the average amortization time for buses up to twenty-four seats is 2.4 years. The duration of the amortization time increases with the size of the bus: for buses from twenty-five to thirty-four seats the duration is assessed at 2.7 years; and for the largest buses in Dar es Salaam, from forty-six seats upward, amortization time doubles to five years (SUMATRA 2011: 41).

These data suggest that business conditions do not allow for exploitation of economies of scale. At the same time, the data do not include key information such as the age and mileage of the vehicles, as well as the years of business experience of the owners—arguably important factors in influencing profit margins. Furthermore, the fact that in 2004 and 2011 the amortization time was the same (at 2.4 years) for small buses illustrates the difficulties in making substantive claims on profitability of the sector. The 2004 study presents no information of what was the size of the bus on which it calculated the amortization time. Furthermore, even assuming one is comparing like with like, there is no sense of how the increasing traffic congestion in the city throughout the 2000s affected profitability of the business and took its toll on amortization.

Thinking about profitability in terms of averages, even when disaggregating by size of buses, has its own limitations. This can be seen in the same 2011 SUMATRA study, when it presents information about owners' own perceptions of profitability. While 71.43 per cent 'of bus owners ranked profitability as moderate', about 20 per cent of the sample ranked the profitability as 'very good' or 'good'. Intriguingly, one reads that this latter group shared the characteristic of having 'supervision mechanism in place'. Frustratingly, no information is presented on what these mechanisms of effective supervision were, as this was 'not within the scope of the study'.

Another example of the tendency to treat owners of *daladala* as an undifferentiated group is the claim, unsubstantiated, that *daladala* business is a 'relay business', with a rapid turnover of bus owners, most of whom go 'bankrupt after three or four years or sell their buses at a loss' (Kanyama et al. 2004: 73). While this insight might apply to some owners, there is evidence to the contrary: some owners do last in the sector. The 2011 study compared the fleet size owned by investors when they entered the business and at the time of the study, and it found that the number of owners with only one bus dropped from 83.8 per cent to 42.4 per cent. By contrast, the number of owners with two, three, four, five, and six or more buses rose, from 12 to 26 per cent for two buses, from 2 to 9.1 per cent for three buses, and from 0 to 10.1 per cent for six or more buses. Taken together, these findings give an indication of the slow, but nonetheless observable, process of concentration of bus ownership. At the time of entering the business, over 80 per cent of owners had one bus, while at the time of the study owners with two or more buses represented 57.6 per cent of the sample (SUMATRA 2011: 36). So the sector worked as a short-lived investment for some owners, who were then forced to pass the baton to other investors and exit, while other owners seemed to be able to slowly accumulate. What differentiated the two groups?

It has been argued instead that behind the rapid turnover lies the fact that the business represents a quick, and therefore ideal, money-laundering mechanism (IDL 2013: 18). While this unverifiable claim might apply to some cases, it is not the whole story. A recurring trope in the literature is the lack of transport professionalism and know-how by many investors (National Institute of Transport 2010: 69). Arguably, knowledge of vehicle mechanics and servicing and, more broadly, of the business is key to explaining why some owners tend to consolidate the number of years in service and fleet size while others do not. As one such owner explained: 'there are two groups of people who own *daladala*. Those who have more than one vehicle and they succeed to add more to their fleet, and those who get money, they buy a *daladala*, they think it will grow money quickly, as they see that every day you get 100,000 shillings, so that gives them hope, as there are passengers at the stops every day' (interview with Mayao, 2010). But all it takes is the purchase of a vehicle in poor condition, or with structural faults that were not spotted, and these owners will face serious mechanical problems and hefty repair sums, that can catch owners out of pocket and, due to most owners being self-funded, unable to continue. Unfortunately, no data allow us to assess how many owners are able to last, and how many belong to the 'relay' team.

In sum, there are serious methodological challenges that stand in the way of researching the profile, profit, and operational realities of bus owners. Above all, they stem from the fact that the real identity of owners is not easily established and that any information about the profitability of the business is politically sensitive. Reviewing existing grey literature on bus owners, their profile, and business conditions therefore requires caution and avoidance of taking its claims at face value. Having said this, there are certainly valuable insights that can be derived from this literature. First, owners are not an undifferentiated group, either by background, source of funding for investment in the sector, or dynamism and durability within the sector. Second, ownership of buses is not significantly concentrated. As corroborated by direct observation in Dar es Salaam, no employer controls enough market share to act oligopolistically. Third, less than 10 per cent of bus owners operate their own bus. Self-employment is thus the exception rather than the norm in this sector. Over 90 per cent of the bus workforce is made up of casual workers who operate buses they do not own. Grasping how the *daladala* sector works therefore requires an understanding of their informal wage relationships, to which the analysis now turns.

3.6 *Daladala* Workers

Kazi mbaya; ukiwa nayo! ('Bad job; if you have one!') (see the photo in Figure 3.2); *Usicheze mbali, unga robo* ('Don't play too far away! There isn't enough flour to prepare *ugali*'); *Maji mengi, unga kidogo* ('Lots of water,

Figure 3.2 *Kazi mbaya; ukiwa nayo* ('Bad job; if you have one!')

little flour').[10] With these words, painted on the back of *daladala*, workers overtly comment on the state of the labour market and on the way in which the oversupply of job seekers is a major problem experienced by workers. Writing about capitalist development in Britain during the industrial revolution, Marx noted the centrality of a reserve army of labour to the accumulation of capital, as it is 'the background against which the law of the demand and supply of labour does its work' (Marx 1976/1867: 790). By 'its work' Marx meant that the existence of an oversupply of workers not only provided employers with additional workers when capitalist production expanded, and with it demand for labourers, but also acted as a depressant on wage levels. While these observations were made about nineteen-century Britain, other work on urban labour in contemporary Africa has noted the oversupply of workers and its negative consequences for workers' employment (Boampong 2010; Lourenco-Lindell 2002). In the introductory chapter, I showed that Tanzania has experienced impressive rates of economic growth since the turn of the century, and yet that this growth has been jobless, with pernicious consequences on its impact on poverty reduction. This chapter now looks at

---

[10] *Ugali* is the main staple in the Tanzanian diet, like rice in much of Asia and pasta in Italy. It is prepared by mixing water and flour, in the same quantities. The mention of 'one quarter of flour' (*unga robo*) or of 'a lot of water' (*maji mengi*), refers to the practice, adopted by the poor at times of hardship, of preparing a watered-down and less nutritious *ugali*.

the consequences of this jobless growth from the vantage point of the labour market for public transport buses, one in which the oversupply of workers affects their room for manoeuvre in negotiating informal labour markets.

The relative ease of entry into the *daladala* labour market stems from the low level of education required. In 2003, survey results found that 78.73 per cent of drivers were schooled at primary level, 16.1 per cent reached form IV at secondary, and less than 1 per cent beyond that. 3.8 per cent of drivers were illiterate. Similarly, the majority of conductors, 76.03 per cent, had only primary education. The percentage of illiterate workers was higher among conductors, at 14.33 per cent (UWAMADAR 2003: 10). The low level of educational requirements for *daladala* workers is common knowledge in Dar es Salaam, as indicated in this ironic advert quoted from a whimsical Swahili article by a local journalist (*Majira*, 18 January 1998):

> *Daladala* conductor wanted. He has to have studied until the seventh level, if not until the fourth level. If not provided with a fourth level class, he has to be able to count money and to give back to the passengers the right change.

The irony encompasses the fundamental fact that any able-bodied man can potentially become a *daladala* conductor. The qualifications necessary to become a *daladala* driver apparently present a greater challenge, with an initial capital of 50,000 shillings being needed to acquire a driving licence in 1998. However, fake driving licences could be easily bought in Dar es Salaam for one tenth of this sum.[11]

What type of workers end up earning a living from *daladala*? The potted employment histories of a few of them, presented below, provide glimpses of the different occupational trajectory of its workforce, and the dynamics that drove their entry into this segment of the labour market.[12]

3.6.1 *Juma Masuka*

Masuka was 32 in 2014.[13] Prior to employment on *daladala*, which he entered as a teenager over twenty years earlier, Juma was a struggling street vendor, trading in sugar cane, a very small-scale business which he abandoned to join a friend who was working as a conductor on a bus. They worked together for a short while, Juma at the door while his friend collected the fares. Another

---

[11] The purchase of fake driving licences has been greatly complicated by the computerization of the system for the award of driving licences, as opposed to the paper driving licences that were in use up until the mid-2000s.

[12] Chapter 6 will present how and whether the careers of some of these men progressed following their employment on *daladala*.

[13] The age of all the workers, whose potted occupational histories prior to work on *daladala* are presented in this chapter, is calculated as at 2014.

friend, who was a *daladala* driver, asked Juma to work as his conductor on a bus called Masuka Trans, from which he took his name and his first full-time job in the sector.

3.6.2 *Kudo Boy*

Kudo Boy was 33 and was originally from the region of Rufiji, in Southern Tanzania. Having completed primary school, and being unable to progress to secondary due to financial reasons, he went to Dar es Salaam at the age of 16 to 'look for a livelihood' (*kutafuta maisha*). Similar to Juma Masuka, Kudo Boy's first economic activity was very small-scale trade, in his case in cosmetic products for women. Like Juma Masuka, Kudo Boy abandoned his business as it provided no profit due to the very small capital base. His connection to *daladala* comes through family relations. His uncle, Kudo Mtalaam ('Kudo the expert', from whom Kudo Boy earns his nickname), a *daladala* driver on route X, provided him with an exit option when he chose him as his conductor in 1998.[14]

3.6.3 *Uwazi*

Uwazi is not his real name, but rather a nickname taken from the title of a Tanzanian newspaper, by which he is known by people at work, due to his previous occupation as a local newspaper seller. While selling newspapers, and during periods of school closure, Uwazi attempted to expand the volume of his business by distributing newspapers to students, who would then work as itinerant newspapers sellers, in exchange for a small commission at the end of business. But as he recalls, when schools opened one year, five of the twenty or so students who were working for him disappeared with all the proceedings from sales, and his capital was eaten away. As Uwazi's kiosk was at a *daladala* station, and he was there very early in the morning for his business, he bounced back from this setback by beginning to work *day waka* shifts as a conductor. 'If someone was late, I would step in and do the work' (interview with Uwazi, 2014).

3.6.4 *Kajembe*

If 'young and failed small-scale trader' is one of the profiles of those who enter *daladala* work, the other main profile is that of much older men, who were drivers, or employees previously, often in the public sector, before becoming

---

[14] All workers whose employment history is reviewed in this chapter worked on the same bus route in Dar es Salaam. To ensure their anonymity, this route will be referred to as route X throughout the book. See also section 4.5 for an analysis of the activities and goals around which workers of this route organized.

daladalamen. Take, for example, Mr Kajembe, a 62-year-old man from Mwanza, in Northern Tanzania, who moved to Dar es Salaam in 1973. He completed primary school and, once in Dar, took a vocational training course in welding. This opened the door for his employment in a garage. In 1986 he became a public-sector employee, as the personal driver of a highly ranked civil servant in the Department of News. In 1991 he was retrenched, and remained jobless until 1995, when the person for whom he was a driver in the public sector bought a *daladala* and employed him as a driver. In twenty years he has worked on several routes, to the point that 'some owners became used to me, they know me' (interview with Kajembe, 2014).

3.6.5 *Rajabu*

Rajabu was 38 in 2014. Prior to employment on *daladala* he was working as a personal driver. When he found himself jobless in 1998, he began working as a *daladala* driver. His connection to his first employer came through his uncle, who took him to a friend of his: 'my nephew is jobless'. The uncle's friend offered him work on a bus, which was operating on route X. After four or five months that job came to an end, as his employer was unable to pay for the maintenance and repairs of the bus, which he sold. By then Rajabu had developed some track record of work on *daladala*. 'I had already done this job, and people knew that' (interview with Rajabu, 2014). One of his neighbours became his next employer: 'I have a bus, interested?' In taking on the job, Rajabu also convinced the owner to shift the route on which his bus operated to route X.

3.6.6 *Rama*

Rama Pengo was 46 at the time of fieldwork in 2014, when he was working as a lorry driver. He is from Dodoma and was educated up to primary school level. He moved to Dar es Salaam at the age of 19, in 1989, in search of work. An imposing figure, his first job in the city was as a guard for a night security company. In 1996 he switched to work as a *daladala* conductor, as a friend of his, a driver on a route X bus, needed a partner. When his friend lost his job following a 'misunderstanding' with the bus owner, Rama was on the bench until another job as conductor materialized, with another driver with whom he had become close while working on route X. That job came to an end when he had to make way for the driver's brother, who had moved from his village to Dar es Salaam to work. After a lengthy spell on the bench, Rama was called to work with Tolu, another driver from route X, with whom he formed a long-lasting working partnership.

3.6.7 *Asenga*

In 2014 Asenga was a 44-year-old man from Rombo in North-Western Tanzania, married with two children. Asenga completed his primary education and, having failed to progress to secondary education, took a two-year mechanics course. His efforts to find a job as a mechanic in Rombo, Tanga, and Dar es Salaam, where he moved in 1988, felt like 'wasting time'. Instead, he started a very small-scale business in second-hand clothes, first on his own, then in partnership with his brother, and having fallen out with him, on his own again. As two years on the business was not going well, Asenga abandoned it and became a *daladala* conductor instead, invited by a friend of his who was a driver.

The employment trajectories of these men into work on *daladala* are diverse and yet underpinned by commonalities. Some men entered *daladala* work as conductors, at a very young age, following unsuccessful experiences as self-employed small-scale street traders. Others, who came from wage employment, for example as drivers, began work on *daladala* having lost their jobs. Still more had quit their jobs because work on buses was a better prospect for them. Notwithstanding this variety of trajectories, employment on the buses therefore acted as a sponge within an economy which was not creating enough jobs: the relative ease of entry into the sector—mediated by personal connections with friends and relatives as well previous line managers—meant that it absorbed the swelling ranks of those seeking employment. But such capacity to absorb workers had its downside, as the oversupply of workers tilted the balance of power with employers decisively in favour of the latter.

3.7 The Employment Relationship in *Daladala*

The position of strength of employers vis-à-vis employees deriving from the ease of entry into this sector's labour market is visible in that, of those workers surveyed in 1998, 82.9 per cent were employed without a written contract and without a fixed wage. In 2003 the figure of workers without a contract was even higher, at 97.7 per cent. The exploitative nature of employer/employee relations is not formalized in any 'agreement' on the workers' wages or conditions of service in the workplace. Workers pay a daily rental fee (*hesabu* in Swahili) to bus owners. The daily return for workers will consist of whatever remains after paying the daily rent to bus owners, petrol costs, and any other work-related expenditure (such as the cost of repairing a tyre or bribing traffic police) has been deducted from the gross daily income. Days in which workers earn no return or even work at a loss are common. When workers end the working day without having enough cash to fill the tank full of petrol, they fill part of the tank. This means that the following day their expenditure on petrol will be higher and their daily return is likely to be lower.

The exploitative nature of the employment relationship, therefore, lies in the 'rental' cost set each day. In theory, this fee is negotiable. In practice, the scope that employees have to bargain is severely curtailed by the nature of the labour market. A driver temporarily 'on the bench' explained the predicament to me in graphic terms (interview with Mashaka, 1998):[15]

> As too many of us are jobless, if for instance a bus owner is looking for a driver, he will find more than fifty people just at this station (i.e. Mwenge station). That is why they can ask you whatever they want and you have to accept it. I worked with the same bus for two years. He used to ask me for 50,000 shillings every day. Over time the buses became too many and the chance of making money decreased. I went to my employer and I told him 50,000 was not possible anymore. He could not understand me and he wanted his bus keys back. He gave the bus to somebody else and he is still working with it. I do not know if he manages to give him back 50,000 every day.

In these circumstances, bus owners can 'squeeze' the fees charged to employees upwards, so that to make ends meet, drivers and conductors are pressed to make a greater number of journeys and to carry more passengers. The magnitude of the worker's daily return must inevitably depend on these two factors. The number of journeys per day is determined by the speed of the driver in negotiating the city traffic, and the number of hours of daily work. The principal limiting factor on the length of the working day is the demand for transportation, a fact confirmed in uniform responses to the questionnaire indicating that the average working day lasts fifteen hours. If the working day is fixed by diminishing demand, then speed of travel remains the crucial variable. Under these circumstances, as another driver explained, the link between workers' exploitation and the high number of road accidents becomes clear (interview with Ramadhan, 1998):

> My owner wants 30,000[16] shillings every day. If I drive without speeding I will work for the whole day to gain only the money the owner wants back at the end of the day. For these reasons we are forced to speed from 5 a.m. to 9 p.m. Then they say too many accidents, how much energy should we have?

While for Marx the length of the working day was the 'product of a protracted civil war . . . between the capitalist class and working class' (Marx 1976/1867), in Dar es Salaam bus owners have such an edge in that war that the length of the working day is not even known: *Punda aumie, mzigo ufike* ('The donkey will

---

[15] See section 4.4.2 for an analysis of the role of workers on the bench in bus public transport.

[16] As the reader will have noted, this figure differs substantially from the previous one. However, what differs is the size of the bus operated, not the amount of work demanded.

get hurt, but its load will reach its destination'), one of Ramadhan's colleagues wrote on his bus. Extremely long working days (the average day lasting fifteen hours and the working week lasting more than 6.5 days) and occupational uncertainty (on average, employment on a bus lasts no more than seven months) are consequences of the very high daily rent that owners expect from bus workers as a result of the power asymmetry between them. Furthermore, the length and hardship of the working day goes a long way in explaining why the workforce is by and large young: 66.3 per cent are between 20 and 35, those between 35 and 45 are 27 per cent of the workforce, and only 6.7 per cent were older than 45 (UWAMADAR 2003: 100).[17]

Workers respond to this financial squeeze by speeding, overloading the buses, and denying boarding to passengers entitled to social fares, all actions that aim at maximizing returns from work on a given day. The analysis of the *daladala* labour market is, therefore, fundamental to an understanding of private-sector performance and the social cost of the 'invisible' hand of the market. The lack of regulation of service by the private sector, and in particular its employment relations, are the key cause of the many inefficiencies of the (privatized and deregulated) transport sector. The words 'Life is War', on the flap at the back of one bus (see Figure 3.1), vividly suggest that workers' unruly conduct has to be understood as a necessary part of the struggle to make ends meet.[18]

Returns from work are highly volatile: over 67 per cent of sampled workers answered 'it depends' to a question on the daily return from work in the 1998 survey. The minority of workers who answered the question indicated an average daily earning of 2,654 shillings in 1998 (approximately US$3.90). Taking into account the cost of living and the existence of dependants, this sum is hardly enough to get by.[19] The texts that workers choose to display in the windscreens of their buses often unequivocally reflect this hardship and the mismatch between their efforts and returns. Among the numerous examples are 'Money Torture' (see Figure 3.3), *Maisha ni Kuhangaika* ('Life is about suffering'), and *Kula Tutakula Lakini Tutachelewa* ('We'll eat, but we'll eat late'). Their situation is similar to Breman's description of rickshaw runners in Calcutta, who also pay a daily rental fee to rickshaw owners and face uncertain daily returns from work. These workers, Breman argued, are not

---

[17] The youngest age bracket for the 2003 questionnaire, from 20–35, is misleading, as many conductors, and a smaller number of drivers, are visibly younger than 20. One can reasonably assume that workers under 20 were 'pushed' into the 20–35 age bracket.

[18] For a similar claim, with reference to Santiago in Chile, see Tomic and Trumper (2005: 56). See also Khosa (1994), on the South African 'taxi' (i.e. minibus) sector and its labour relations in the 1980s and early 1990s. While there are important contextual specificities, the uneven bargaining power between bus owners and workers is remarkably similar.

[19] The 2003 survey contains no information on daily returns, arguably due to how unpredictable they are.

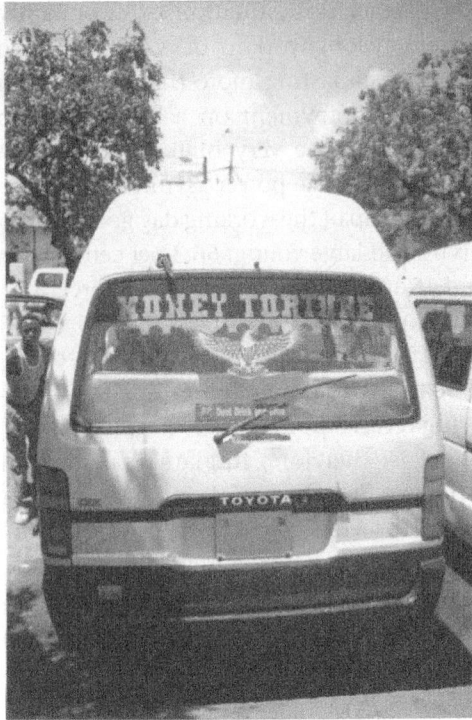

Figure 3.3 'Money Torture'

'independently-operating small entrepreneurs, ... but dependent proletarians who live on the defensive' (Breman 2003: 154).

The modalities of employment and remuneration of the workforce can be best understood as a strategy by bus owners, or de facto employers, to transfer business risks squarely onto the workforce. Bus owners confront labour not as risk-taking entrepreneurs, but as rentiers, leaving workers to manage the risks inherent in low and volatile productivity, a condition that is conducive to self-exploitation by the worker rather than to growth in productivity. In these circumstances, therefore, workers act as entrepreneurs only in the sense that they have become managers of two sets of risks under adverse conditions of extreme competition: the daily insecurity that results from an uncertain income, on the one hand, and the ever-present chance of erratic job loss, on the other (Wuyts 2011).

These are the informal employment relations that prevail in the private buses that provide public transport in Dar es Salaam. These workers do not earn fixed wages, nor are they pieceworkers. That workers are not waged in a conventional sense does not imply, however, that labelling them as self-employed micro-entrepreneurs, as policymakers and official statistics on the informal economy commonly do, is a better fit. Instead, their case illustrates the way in which conventional categories of both 'wage/paid employment' and 'self-employment' do not easily apply to the reality faced by informal workers and the complexity of the employment relationship that links them to employers. At the same time, however, it is important not to lose sight of two key characteristics that define their employment status and, ultimately, the economic stakes held by workers in the labour process. First, these workers do not own any of the capital with which they work. A clear division between capital and labour can be observed here, making the notion of self-employment implausible in this case, as it would convey a misleading notion of entrepreneurship and economic independence and conceal the fundamental power relation at play between bus owners and workers. Second, it is precisely because of workers' economic vulnerability that they are deprived of a conventional wage employment relationship with employers. Theirs is very much 'a disguised form of wage labour which deprives workers even of the meaning of a proletarian work relationship' (Castells and Portes 1989: 13).

Importantly, *daladala* workers see themselves as casual wage workers rather than as self-employed workers, as is evident from the goals of their political organization since the late 1990s, the focus of Chapter 5. As we will see, when these workers established their association, and built an alliance with the Tanzanian transport union, their main objective was to lobby the state and employers to spell out the employment relationship with bus owners. Their agenda was no entrepreneurial agenda, nor did it entail a request for micro-credit. Instead, they demanded employment contracts and a fixed wage. Such characteristics of workers, and their own perception of their economic interests, exposes the analytical limitations of framing the study of economic informality in terms of 'the conflict between the government and those involved in the informal economy' (Tripp 1997: 136), with its glossing over of the importance of class relations and other forms of socio-economic stratification, especially within the informal economy itself. In the case of the bus public transport system, the conflicts of interest between workers and their employers are a crucial dynamic in understanding how *daladala* operations work.

As such, *daladala* workers are best categorized as people in paid employment in the informal economy. Many other forms of paid employment are to be found in different economic sectors and in different contexts, with the working poor often combining precarious wage employment with some ownership of equally insecure, very small-scale activities in the informal economy (Bernstein 2010). Notwithstanding the heterogeneity of such employment relations, however, they do not easily match the conventional conceptualization of either 'paid employment' or 'self-employment'. Yet labour force surveys consistently suggest that paid employment is statistically irrelevant in the informal economy, and self-employment the norm. Leaving the streets of Dar es Salaam behind for a moment, the analysis now looks at how wage employment and self-employment are defined in questionnaires, and then translated into Swahili, in order to generate statistics on the informal economy.

3.8 The 2006 ILFS Questionnaire and Informal Wage Employment: Lost in Translation?

The importance of paying attention to the way in which key employment and work concepts are worded by labour force surveys is well understood (Campanelli et al. 1989; Martin and Polivka 1995). However, the way in which concepts are translated from English into other languages in labour force surveys is less often investigated (Floro and Komatsu 2011). This is problematic as ultimately it is in languages other than English that questions are posed to labour force survey respondents. Translating words and concepts, often ideologically loaded and context-specific in their origin, into other languages is not an easy task. Much can be lost in the process of translating the labour force questionnaire into Swahili. Moreover, questionnaire respondents make sense of employment questions in light of how they perceive the realities that they confront in their daily working lives.

Notwithstanding concerns about the household as a unit of analysis (Cramer et al. 2014; Guyer and Peters 1987; Randall and Coast 2015), consider, for instance, the ILFS introductory question on household economic activities, to be answered by the head of the household on behalf of his/her household members. In English it reads:

> Does this household or anyone in this household engage in any of the following activities? a) Wage Employment (yes/no); b) Working on own or family business (excl. Agriculture); c) working on own *shamba*, fishing or animal keeping; d) do you have any paid employees?   (NBS 2009a: 3)

What differentiates the four (not mutually exclusive) possible answers, at least in the English version of the questionnaire, are the three possible types of employment status: (1) being a wage employee; (2) being self-employed; and (3) being an employer. In Swahili, however, 'working on own or family business' is translated as *kazi isiyo ya kilimo* (NBS 2009c), which literally means any 'work that is not agriculture'. This means that the reference to self- or family employment in business or agriculture, central to the English wording of the questionnaire, is dropped altogether in the Swahili version.

The section of the questionnaire on the individual respondent's main economic activity (rather than on households at an aggregate level) does better, as it presents an accurate correspondence between English and Swahili survey questions. Respondents are asked whether their work entails self-employment: *kujiajiri mwenyewe binafsi*, which literally means 'to hire oneself'. But what most influences respondents' choice of the category that best describes their employment status is how they understand the main alternative answer they might opt for, namely 'paid employment'.

In the 2006 ILFS, the Swahili translation of the term 'wage employment' is also problematic. The term used in this case is *ajira ya msharara*. While this literally means wage (*mshahara*) employment (*ajira*), such terminology clearly connotes registered employment in the formal sector, 'proper jobs' for the lay Swahili mother-tongue speaker, with a formal employer, a contract, and a wage. Part of the problem lies in the ambiguity of the term *ajira* in Swahili. Broadly speaking, *ajira* is used to denote employment of any type. In this sense, one reads and hears that *Tanzania tatizo ni kwamba hakuna ajira* ('The problem in Tanzania is that there is *no employment*') (interview with Dongo, 2010). Yet, at the same time, people use the word *ajira* to mean registered employment, as opposed to employment of precarious and informal nature. Along these lines, a *daladala* worker, commenting on his informal and precarious employment, stated that *Tatizo la kazi ya daladala ni kwamba hakuna ajira. Kibarua tu* ('The problem of work in *daladala* is that *there is no formal employment*. Only casual work') (interview with Ngaika, 2009). Failure to appreciate the two possible meanings of the word *ajira* in Swahili would potentially allow the implausible translation of the sentence above as 'The problem of work in *daladala* is that there is no employment'! Instead, when workers refer to their work as work without *ajira*, they mean that it is work without contract, rights, and security, in other words, informal. This suggests that there is therefore a remarkable dissonance between the way in which the concept of 'paid employment' is worded in Swahili by ILFS, and the way in which informal casual workers understand and word it.

The bias against recording informal wage employment is present also in the questionnaire section focusing on working patterns of individual members of the household. The question on 'what was the economic activity in which you spent most of your time?' has 'employee in a wage job' as one of its five possible answers (the other four being self-employed, working on your own or family farm, unpaid work in family business, and other). The Swahili wording of 'employee in a wage job' as *mwajiriwa wa kulipwa* once more points to formal-sector employment. And so does the range of subsectors in which an 'employee in a wage job' might be employed: the central government, the local government, a parastatal organization, a political party, cooperatives, NGOs, international organizations, religious organizations, and the private sector (NBS 2009b: 3). It is highly plausible that a respondent answering this question will fail to match his/her informal employer with any of the possible employers from the survey list, and will not opt for declaring himself/herself as an 'employee in a wage job'.

ILFS, therefore, puts forward a stark and questionable dichotomy between paid and self-employment, and a leading one at that. Consider the implications of the translation issues of 'self-employment' and 'paid employment' together. On the one hand, 'self-employment' is translated in extremely loose terms, to the point that any work outside agriculture seemingly fits into it, or that work by people who do not own any capital can be misleadingly identified as 'self-employment'. On the other hand, paid employment is translated in very narrow terms, so that only those in formal and registered paid employment are likely to identify themselves as 'paid employees'. Arguably, it is out of this contrast between an overly expansive notion of self-employment and an extremely narrow notion of paid employment that the official statistics are created, thus suggesting that the informal economy consists of a teeming mass of family entrepreneurs.

A depiction of economic informality as self-employment is then consistently built upon by the 2006 ILFS, specifically through its modules on the informal economy, where information on informal business is sought (see NBS 2009b: 8, questions 26–32, which are designed for 'business owners only'). The focus is on understanding how businessmen in the informal economy set up their businesses, from where they operate and how often, and their sources of credit and training, but without much consideration of how many of these respondents can really be understood as businessmen in any meaningful way.

3.9 Informal Wage Employment: Invisible and yet Central

This chapter had two related goals. First, it has documented the significance of socio-economic differentiation and of class in bus public transport in Dar es Salaam and its centrality in understanding how *daladala* operations work. Second, as informal wage employment hardly exists according to official statistics on the informal economy, the chapter has interrogated how such statistics are created. The analysis has shown that the importance of wage employment has been largely rendered invisible, in part because wage labour is often overlooked and lumped together with self-employment, and in part because the significant trend towards the subcontracting of informal labour services rather than the direct production of commodities is poorly understood. For example, while the *daladala* workers in urban Dar es Salaam are directly involved in the sale of urban transportation services, it is nevertheless questionable to classify them as self-employed given the lack of control they have over the capital that they operate and the precarious nature of the terms of their 'employment'.

At the root of the invisibility of informal wage labour lies the fact that conventional categories of 'self-employment' and 'wage employment', on which labour force surveys rest, are inadequate for capturing the heterogeneity of employment relations found in the informal economy and the heterogeneity of relationships between capital and labour that mediate poor people's participation in the (informal) economy. Using the case of Tanzania, this chapter has highlighted the remarkable distance between the complexity of the employment relationships linking informal wage workers to employers and the clear-cut categories used to frame questions for the 2006 ILFS. As field-based research repeatedly shows, the simple dichotomy between self-employment and wage employment does not work. More attention needs to be paid in survey analysis to the results of field-based qualitative research that does not only focus on differences in earnings, but also on the myriad of labour regimes that prevail in informal production.

The analysis has further argued that the Swahili words chosen for asking workers whether they are in wage employment communicated a very narrow connotation of paid employment in the formal sector. By contrast, 'self-employment' is translated in extremely loose terms, arguably acting as a 'catch-all' category in the Tanzanian context. The ILFS statistical suggestion that only 0.7 per cent of workers in Tanzania's informal economy are wage workers, and the remaining are self-employed in one way or another, therefore, rests on disturbingly shaky grounds. To address this major shortcoming requires in-depth research to understand the language and categories used by informal workers. Such research would be essential to design better survey questions, aimed at detecting and understanding the nature of informal wage labour, or the work of *kibarua*, a word ubiquitously referred to by informal wage workers in Tanzania to describe their status, yet a status that is strikingly at the margin of the 2006 ILFS. If the picture of informal economies presented by ILFS has indeed no analytical purchase on actual realities on the ground, as we would argue with reference to the Tanzania 2006 ILFS, efforts to identify labour categories that are intelligible to respondents should take priority.

The chapter has also emphasized the urgent need to move away from the problem of 'misplaced aggregation' in the classification of labour regimes, which results from conflating into one catch-all category various forms of production and employment that are essentially different, not just as static entities, but also in terms of their dynamic potential. It is difficult to see how one can address the issue of the dynamic potential of the informal economy without taking explicit account of these diversities in production and their corresponding labour regimes. Coming to terms with these issues, however, would require a shift in focus towards the analysis of capital accumulation and its relation to the transformation of labour regimes in the so-called informal economy, an issue on which mainstream literature on economic informality is sorely silent and labour force surveys provide insufficient insights.

The chapter has begun to put to work such a research agenda in the context of bus public transport in Dar es Salaam. Through its focus on labour relations in the transport sector, it has analysed the costs of the lack of regulation of employment relations and its causal link to the many inefficiencies of bus public transport in the city. The oversupply of unskilled workers significantly tilts the balance of power towards bus owners in the 'negotiation' over terms and conditions of employment. In contrast to the fashionable emphasis on the agency of the urban poor in developing countries, a claim often made without due attention to the formidable structural constraints they face, the chapter has shown that the most tangible way in which these highly vulnerable and exploited workers exert their agency is by dragging society into the 'war' that work as a *daladalaman* entails. A war which consists of speeding, the systematic negation of road safety rules, and other such 'guerrilla' tactics.

4

The Politics of Labour 1

The Quiescent Period (up to 1997)

4.1 The Criminalization of the Workforce

In previous chapters, I have suggested that the lack of regulation of service by the private sector, and in particular of its employment relations, is the key cause of the many inefficiencies experienced by passengers of bus public transport in Dar es Salaam. In a sense, theirs is a form of referred pain from that of the workers. The unruly conduct by the workforce has to be understood as a necessary part of the struggle to make ends meet and above all to preserve their (casual) employment in the face of employers' requests for a hefty daily sum. However, this is not how Tanzanian institutions, the public sector, the association of bus owners, or even the general public frame, explain, and respond to the problems with the transport system. Over the years, a recurring narrative is visible, that criminalizes the workers and attributes the many accidents and the chronic tensions within the transport system to the hooliganism and greed of its workforce. Newsprint frequently refers to the need for a 'cleaning' of the transport system. Cleaning proposals and practices take more or less authoritarian forms. For example, requiring the workforce to wear identification badges and uniforms, first introduced in 1991 by the Minister of Home Affairs to facilitate control of workers refusing to ferry students, is a measure that the state has adopted time and again to discipline the workforce, each time with no tangible results.[1] A more authoritarian version of attempts to bring order to the transport system was the imprisonment of 'dirty workers', an experience shared by so many that prison has earned itself the nickname 'university'—i.e. where lasting lessons are taught.

---

[1] See *Daily News*, 28 June 1991, for the first time; *Mtanzania*, 2 September 2007, for a more recent recurrence.

Thus, even in the case of *daladala* workers and of public transport problems, the Tanzanian state understands 'the causes of urban poverty in terms of the poor themselves. Poverty is not seen as a reflection of the organization of the entire socioeconomic and political system. Because of this, the solutions instituted by the state to solve urban poverty end up treating the phenomenal forms rather than the essential relations behind the problem' (Lugalla 1995: 178–9). The reasons for such a myopic approach need interrogation, starting with the observation that events in Dar es Salaam show striking continuity with a tradition of criminalization of the urban workforce and the unemployed by the state in colonial Africa (on colonial Dar es Salaam, see Burton 2005; on sub-Saharan Africa, see Cooper 1983 and Worger 1983) and more recently across the developed and developing world (Wacquant 2003, 2010). As Bernstein notes (2010: 116), 'the economic and social power of capital, rooted in a system of property and commodity relations, has to be secured through its political and ideological rule, exercised . . . through the state'. In this particular case, the state, whose employees include bus owners, frames transport problems in a way that ideologically—and not accidentally—downplays the importance of unregulated relations of employment on the behaviour of workers.[2]

Why did workers accept this state of affairs? A major absence is noticeable here, namely that of the voice of the workers themselves. Interviews during fieldwork in 1998 and 2001–2 elicited a strong feeling of anger towards the lack of intervention by the state with regard to the fee expected by employers, and with employers' lack of willingness to understand workers' circumstances on days in which the sum of money expected from them could not be collected. Workers also expressed profound disillusionment with the integrity of the government's various attempts 'to clean' the transport system. Such a feeling is effectively illustrated by a *daladala* worker through the following words, displayed on a *daladala* windscreen: *Kama kuoga ni usafi kwa nini taulo imechafuka?* ('If by washing one gets clean, why does the towel get dirty?'). However, beyond disillusioned sarcasm, up to the mid-2000s transport workers had not engaged collectively with public institutions or employers over the terms of employment and/or the direction of transport policy more broadly.

---

[2] It should be noted that the attitude of the state towards transport problems is not always functional to the interests of bus owners, as shown by another instance of tackling a symptom (speeding) rather than the cause of transport problems. On 31 October 1996, the Parliament approved an amendment to the 1973 Road Traffic Act, which prevented any vehicle from providing public transport unless a speedometer and a mechanical device limiting speed to 50 km per hour was installed (United Republic of Tanzania 1996: 146). The measure was abandoned five months after its initial enforcement, as it emerged that high-level officers had pushed for the measure with the aim of pocketing kickbacks from the two local firms that were allocated the business of installing the mechanical speed limiters. Urban transport policy thus reflects the outcome of a political struggle among competing interests.

This chapter is the first of two investigating the political sociology of transport workers. Here the focus is on the reasons behind workers' lack of voice and political quiescence, by which I mean their failure to articulate demands to bus owners and the state, as this is what could be observed until 2002. The chapter also analyses the existing forms of solidarity among the workforce, and their significance, notwithstanding their failure to collectively take on their employers and the state.

Later spells of fieldwork took place in a very different political landscape due to the establishment, in 2000, of UWAMADAR, a *daladala* workers' association. The very presence of an institution, founded by transport workers for transport workers, reversed, at least formally, the political asymmetry that had characterized urban transport policymaking to that date. In response to these events, later rounds of fieldwork in 2009, 2011, and 2014, the findings of which are presented in the next chapter, explored the reasons behind workers' more assertive and successful attitude towards the state and employers, in partnership with the Tanzanian transport union. Read together, the two chapters thus underline the importance of historicizing the study of the political organization of informal workers, with its twists and turns, and in its complexities, by allowing time frames that are broad enough to capture the slow pace at which changes in political attitudes unfold.[3] They also show the limitations of reductionist and functionalist readings of capitalist domination vis-à-vis workers. This is not to deny the significance of the uneven balance of power between employers and workers, which we have already discussed and to which I will continue to pay attention. Rather, the point is to emphasize the importance of incorporating into the analysis attention to actual mechanisms of domination and to whether and how such attempts at domination are resisted or not by workers (Cooper 1983: 29).

The chapter draws on the in-depth and longitudinal (1998, 2001–2, 2009, 2010, and 2011) study of one *daladala* station and of one particular route that started from there. Many of the insights that inform the analysis herein presented were generated by 'hanging out' there: by sitting on a bench at one of the station kiosks or at the end of the route, and by observing the interactions of workers. Over time I became 'staff', as I was jokingly referred to by workers, even forcefully being granted exemption from fares when travelling on the buses of that route. Having become acquainted with some of these workers, I was in a position to ask for clarifications about the meaning of actions that I could observe. In this process I came to learn about the existence

---

[3] As testament to the way in which processes that are underway at a given time in a given context might not be easily visible to the researcher working on them, during fieldwork in 2001 and 2002 in Dar es Salaam the association of *daladala* workers had no public visibility as its leaders were engaged in building the partnership with the Tanzania transport trade union, as we will see in Chapter 5. Only in later years did the two institutions begin making their demands public.

of one association, consisting of people from various 'classes of labour' working on one bus route. While they were all earning a living by working on *daladala*, they did so by carrying out different tasks related to public transport. I attended some of the association's meetings, consulted its records (see sections 4.5.1, 4.5.2, and 4.6) and interviewed some of its members. Such ethnography of workers' associationism aided the development of my research along three lines of enquiry.[4] First, the associational goals that *some* workers set for themselves, and the values that these reflected, facilitated not only an understanding of workers' strategies to negotiate precariousness, but also of whether their collective action embodied a moral economy, or an alternative set of values by the urban poor. These findings are therefore to be read in critical dialogue with the narrative of urban Afro-populism, reviewed in Chapter 1, which celebrates the capacity displayed by African cities to generate 'a new urban sociality even under dire conditions' (Simone 2010: 314) and even claims that such practices 'must be the touchstone of radical imaginings and interventions'. I did indeed open my eyes to 'the phenomenology and practices of the "everyday"' (Pieterse 2008: 9) and hereby present an empirically grounded account of what 'ordinary' citizens—never mind how vague such a category is—do in one African city. Second, this close-up study of workers' associationism and, in particular, observation of the daily interaction and tensions between different categories of transport workers across the association's life, allowed exploration of the ways in which labour oversupply, labour fragmentation, and other factors impact on workers' identities, as they erode the workers' common ground. The study of the association's politics therefore opens a window into the 'struggle over class'. Last, but not least, the patterns and rhythm of labourers' circulation *within* the sector are an important element in understanding both the precariousness faced by transport workers and their political mindset. Access to the association's archive, and most notably to its data on the turnover of members, allows some appreciation of the modalities of labour fluidity over time.

4.2 The Sources of Workers' Power

Understanding the reasons behind workers' political quiescence in one particular context must begin with a conceptualization of the sources of workers' power. I will draw on Wright's conceptualization (2000). This is widely adopted (see, among many others, Agarwala and Herring 2008, Kabeer et al. 2013, Selwyn 2007, Silver 2003, Von Holdt and Webster 2008; Webster 2015),

---

[4] The name of the association and the route to which it was linked is kept anonymous to protect informants.

arguably for the clarity with which it sets out the importance of analytical attention to class and the structural location of workers, while not presuming, in a crudely reductionist way, that their bargaining power can be crudely derived by their structural position. According to Wright, workers' power can be schematized as deriving from two possible sources. First is the 'structural power' that (some) workers command. This derives from workers' specific 'location . . . within the economic system'. Following this argument, some economies, and some industries within them, have more potential to generate labour unrest than others. More specifically, there are two subtypes of 'structural power' to be considered. By 'marketplace bargaining power', Wright means the power that workers command due to conditions in the labour market across economic industries. A tight labour market—that is, one in which employers struggle to secure their supply of labour—will lead to workers' high 'marketplace power'. Conversely, an economy experiencing an oversupply of unskilled workers will result in such workers' low 'marketplace power'. The second subtype of 'structural power' is 'workplace bargaining power'. This results (or not) from the specific industrial location of workers, e.g. minibus workers operating in Dar es Salaam, horticulture estate workers in Brazil, call centre workers in Bangalore. Whether workers take advantage of their structural power rests in turn on their second source of power, which is 'associational' rather than 'structural'. This derives from the political organization of workers along trade union lines or other institutional forms. Crucially then, and against any teleology, there is no straightforward correlation 'between workers' bargaining power and the actual use by workers of that power . . . to struggle for better working and living conditions' (Silver 2003: 15). Whether the socio-economic position occupied by workers translates into political consciousness and a shared identity depends on a number of complex factors (see Bernstein 2010: 115–17, for a synthesis). Most notably, workers do not automatically experience (and make sense of) exploitation 'self-evidently or exclusively' along class lines, either in an unmediated or direct fashion. Active efforts (by outsider activists or by workers themselves) at constructing a shared notion of injustice and exploitation play an important role in successful instances of workers' political mobilization.

In light of the above schematization, what are the sources of power that bus transport workers in Dar es Salaam command? *Daladala* workers have low 'marketplace power'. As we have seen, the existence of an oversupply of unskilled job seekers significantly tilts the balance of power between bus owners and bus workers in the former's favour, as the daily rent expected for a day of work is imposed by owners on workers and is non-negotiable. While these workers command low 'marketplace power', the urban bus public transport sector in which they work confers on them, at least potentially, 'workplace power' (the other subtype of structural power). As Silver noted, 'transportation workers have had, and continue to have, a strategic position within the world capitalist economy and within the labour movement' (Silver 2003: xv). Such an empirical finding is explained not so much by the 'direct impact of their actions on (often public) employers' but rather by the 'upstream/downstream impact of the failure to deliver goods, services, and people to their destination'. Such insight arguably applies to the Dar es Salaam case. As private buses have long constituted the only means of (barely afford-able) motorized public transport available to the public, unrest by its work-force would seriously affect the mobility of the vast majority of Dar es Salaam's population. However, their inability to take advantage of their structural power was caused by their lack of 'associational power'. This was evident in the absence, up until 1997, of an institution representing workers in urban transport policymaking. What, then, were the barriers militating against the organization of workers?

4.3 The Spatial Unit of Work

The oversupply of unskilled workers, a phenomenon that can be observed across the economy, significantly undermines the bargaining power of *dala-dala* workers. To understand how the mismatch between demand for *daladala* labour and its supply affects workers, attention must be given to the spatial unit in which work is performed. The concentration of workers in one place has been shown to aid their organization and collective action (Iliffe 1979: 399). For example, during Dar es Salaam's famous dockworkers' strikes in the late 1940s, a group of unskilled dockworkers was able to make claims on employers and to win concessions, notwithstanding similar issues of labour oversupply. A different period and under very different conditions, of course, but the point here is to emphasize how part of the dockworkers' strength rested on being able to temporarily close off the docks, preventing other workers from breaking their strike. The atomized nature of the 'workplace' in the bus public transport system, with about 30,000 people working on about 10,000 buses, acts as a barrier to collective action.

4.4 Labour Heterogeneity: The Phenomenology of Transport Workers

There is no such a thing as 'a transport worker'. Appreciating the significance of class fragmentation in the labour market is one of the first steps to under-standing the lack of voice of its workforce. Davis's observation that the infor-mal economy 'generates jobs not by elaborating new divisions of labour, but by fragmenting existing work, and thus subdividing incomes' neatly summarizes what can be observed in Dar es Salaam (Davis 2006: 181). Three different divisions or 'classes of labour' are visible in the transport labour market.

4.4.1 Daladalamen 'with a Livelihood'

The first category is referred to by the workforce as *daladalaman maisha*, with a double connotation to it, *daladala* worker 'with a livelihood'. Such workers usually work in pairs (a driver and a conductor). Despite the wording chosen by workers to identify this group, 'workers for life' are not permanent employees as formal contracts are very rare. In each bus either the driver or the conductor is in charge of the vehicle and, as such, is responsible for returning the daily fee to the bus owner. Notwithstanding this, both driver and conductor share the need to spend the best part of the day, and by aggregation, their 'life' working on the bus. This is not an overstatement considering that the working week (or life) of a *daladalaman* typically consists of fifteen hours per day for an average of 6.67 days a week (Appendix A). One also hears a different connotation, evident when a driver or conductor is referred as *dereva/konda ambaye ana maisha* ('driver/conductor with a livelihood'). While literally *maisha* means 'life', the accurate translation is 'livelihood', as what identifies these workers is that they have a job (for as long as it lasts), through which they can earn a living. This outlines their difference from workers 'on the bench'.

4.4.2 *People 'on the Bench'*

The second category of workers is known either as *day waka* ('day workers') or *watu wa benchi* ('people on the bench'). The distinctive characteristic of this second group is that they are part-time day workers whereas '*daladalamen* with a livelihood' might be termed full-time day workers. The interaction, and difference, between '*daladalamen* with a livelihood' and 'people on the bench' is best understood by drawing on an agricultural metaphor often used by transport workers: *kukomaa* ('to overripen'). '*Daladalamen* with a livelihood' 'overripen' when they operate the bus from the beginning until the end of the working day. This is extremely taxing on workers' physical and mental well-being, considering the heat, the need to speed, to avoid collisions with other speeding buses and impounding by road police, to engage in turf wars with school pupils, and to do all this in the presence of often argumentative passengers. Severe back pain, mental distress, and its somatization in the form of ulcers, are the vicious (and common) manifestations among workers from such extreme working conditions. As one worker put it, 'you cannot overripen in the bus every day . . . you will not feel well. Exhaustion, you start every day at 4 a.m.' (interview with Dongo, 2014). And in the words of another, 'you can overripen for an entire week, then you need to rest at home for three days. I wake the bus up (literally *kuamsha gari*) at 4 a.m., and I am back at home at 9 or 10 p.m., if I am early, and in bed by 12' (interview with Abasi, 2014).

As far as possible, '*daladalamen* with a livelihood' avoid overripening by taking a break for part of the working day. During that time *day waka* take over the operation of the bus. On a fairly bad day this might be for only two trips. Once dismounted, the *day waka* will be paid a small amount, normally enough to buy a cheap meal. On a less bad day, when the '*daladalamen* with a livelihood' need or can afford more than an hour off, the shift of the *day waka* is more substantial, as much as half of the working day, and so is his share of the cake. Workers thus only 'overripen' on bad days, when they predict that their return from labour will be insufficient to divide what they earn with bench workers.[5] Given the shortage of alternative employment opportunities, there are almost equal numbers of workers 'with a livelihood' and *day waka* at any station. This estimate is borne out by the author's fieldwork observations and also by records of the workers' association. The latter records that in December 2001, the number of *daladalamen* 'with a livelihood' on one route was fifty-four while that of workers 'on the bench' was fifty-six.[6]

While the difference in tasks performed by the two classes of transport workers is easy to grasp, the dynamics that lead a person to become a *bench* player need explaining. The vast majority of these fifty-six were former '*daladalamen* with a livelihood' who had lost their employment status for various reasons. Twelve of these workers (21.4 per cent of the sample) had had an argument either with the owner or the driver. For fourteen (25 per cent), the buses they had worked had been sold. For twelve (21.4 per cent), the vehicles had broken down and had not been repaired. Six (10.7 per cent) had been involved in road accidents. Five (8.9 per cent), had fallen ill. The remaining seven *day waka* (13 per cent) were people the association had 'become accustomed to'. A teenager, who at the time of fieldwork in 1998 used to sell water at the station, was part of this group, as were two people introduced to the station by '*daladalamen* with a livelihood'.

---

[5] Religious practices influence the division of labour between '*daladalamen* with a livelihood' and *day wakas*. During Ramadan, fasting Muslim workers will only work for half the day. Interestingly, religious syncretism takes place as non-Muslim workers follow suit. Ramadan is therefore the best month of the year for workers 'on the bench'.

[6] There were two '*daladalamen* with a livelihood' working on each of the twenty-seven buses serving the route in December 2001 (giving a total of fifty-four). The list of the fifty-six 'bench workers' for the same period appeared in an association file, entitled '*Wasio na kazi* ["The jobless"]— Name of Route', a copy of which is in the author's possession.

4.4.3 'Those Who Hit the Tin'

Wapiga debe (literally 'those who hit the tin') is the last category of transport workers. In a congested bus public transport system such as that of Dar es Salaam, at any one time and at any given station, there is more than one bus that can be observed leaving for the same destination. En route the crew solicit prospective passengers, whilst at final stations *wapiga debe* take over this task, encouraging passengers by hitting the body of the bus and shouting its final destination. Once the bus is full, they are paid the price of one fare by the crew. *Wapiga debe* fill up bus after bus throughout the working day. This category of transport 'call boys' appears at most bus stations in urban Africa (on Malawi, see Tambulasi and Kayuni 2008; on Africa more generally, see Godard and Turnier 1992).

Wapiga debe occupy the very bottom rung of the transport workforce ladder, both in terms of social status and the unskilled nature of their job. The main skill they require is the ability to shout, a task that, performed over time, wears out their vocal cords, hence their characteristically cracked voices. Many smoke heroin and marijuana, or sniff glue, to cope with the scorn they are subjected to as a result of the demeaning nature of their work. As one of them put it (interview with Dotto, 2010):

> Any man with a sound brain knows that shouting a destination and pulling people into the bus all day is not a job. We do it because we are in trouble. But it is not a job... The heart hurts when you think about life, because it is not life to be here at the station. You can't bring your family 'Come to the office'... This is a pavement and as it is a pavement it is not an office.

Wapiga debe are the outcasts par excellence of the transport system and the attitude of public institutions towards them is even fiercer than towards other categories of transport workers. Campaigns by municipal authorities to crack down on *wapiga debe* are a regular feature of the transport system, occurring in 1993, 1996, 1997, 2005, 2007, and 2009. In 2005 the Dar es Salaam City Council even hired a private security firm to deal with the problem, although to no effect; when the security firm proceeded to confront the *wapiga debe*, some of its staff members were injured by a group 'aggrieved to lose its source of livelihood' (*The Guardian*, 9 July 2005).[7]

Given the lack of alternative employment opportunities, campaigns to do without *wapiga debe* have proven unsuccessful. Furthermore, and notwithstanding the low social status of the job, competition is fierce among a vast number of *wapiga debe*, their age varying from early teens to forties. *Wapiga*

---

[7] See also *Uhuru*, 15 April 1993; *Nipashe*, 17 February 1995; *Daily News*, 20 March 1996; *Uhuru*, 12 May 1997; *Mwananchi*, 2 September 2007.

debe in most places are organized into gangs, which 'negotiate' solutions to market congestion. Negotiation can take different forms, from petty warfare to a more peaceful division of the market. '*Daladalamen* with a livelihood' often have little choice but to let *wapiga debe* take over part of their work (and income), as they are exposed to varying degrees of intimidation and potential retaliations by gangs. Turning down the help of *wapiga debe* might result in a smashed window or a flat tyre, and bearing the cost of repairs.

The analysis has so far documented the significance of socio-economic differentiation between classes in the transport system and the fragmentation of its workforce between different 'classes of labour'. Although '*daladalamen* with a livelihood', *day wakas*, and *wapiga debe* share the characteristic of selling their labour power in the transport sector to earn a living, the three groups perform different tasks from one another, with differing levels of insecurity and remuneration. In most *daladala* stations, and for most routes, the interaction between the three categories follows the pattern described, one in which overworked workers subcontract part of the day to underemployed workers, and in which gangs of men vie for space at stations to fill buses. On some routes, however, workers interact in a way characterized by a higher degree of organization.

4.5 Workers' Associationism: Forms and Limits of Solidarity

Keeping in mind these 'classes of labour', the analysis now investigates one instance of transport workers' associationism and yields three types of insight. First, exploring the goals that workers set for themselves, and the values that these reflected, is an entry point from which to reflect upon the 'generative powers of the ordinary' as African urban populists would have it. Second, the analysis focuses on the interaction and tensions between different categories of transport workers across the association's life to illustrate the form that the 'struggle over class' took in this context. The key question addressed is the extent to which labour oversupply and labour fragmentation impacted on workers' identity. Last, but not least, access to the association's archive, and most notably its data on the turnover of members over time, allows an appreciation of the patterns and rhythm of labour circulation *within* the sector. As such, it constitutes an important source for understanding both the precariousness faced by transport workers and their political behaviour.

4.5.1 *Managing but not Challenging Precariousness*

The association under scrutiny was formed in 1998. It groups together workers from the three 'classes of labour' in the transport system (i.e. workers 'with a livelihood', 'bench workers', and *wapiga debe*) and does not include bus owners or self-employed bus workers (i.e. those operating their own buses). It was established following a road accident involving a *daladala*, in which the driver was found guilty of serious violations of road safety rules and sent to jail. The owner of the *daladala* decided not to become involved. In a move that signalled collective consciousness of the plight of their colleague, workers amassed enough money to bribe the jail personnel and obtain the worker's release. The initiative per se was not a new step, as being jailed was, and still is, part and parcel of working in the Dar es Salaam transport sector and workers had previously responded in a similar fashion. However, having settled the problem on this occasion, a group of workers from the route decided to shift from ad hoc, yet highly frequent, collections of funds to aid colleagues in trouble, to a more systematic way of saving funds as a cushion against future problems. They were later able to convince others to join them.

Attention to space is key to understanding why transport workers from this particular route were able to institutionalize their interaction in contrast to workers from other routes, and why the organization took the specific form it did. Workers were able to exploit the fact that their route shared no more than a handful of bus stops with other routes. This characteristic is the exception rather than the rule; in Dar es Salaam most bus routes overlap, thus the vast majority of bus stops are shared.[8] Taking advantage of their de facto monopoly of service, workers self-regulated the way in which they provided transport by doing away with the cut-throat competition over fares and passengers that can be observed in unorganized routes. Instead, they formed orderly queues at both ends of the route, so that at any one time there was only one vehicle waiting to fill up before departure, a system to which workers referred to as 'the railway line'.[9] Furthermore, they opted not to compete with each other over fares, instead setting the fare throughout the day at its rush hour value of 150 shillings. This measure yielded substantial additional income to the workforce as, on unorganized routes, the excess of buses forced crews to lower the fare to 100 shillings during off-peak times.

The way in which workers prevented competition over fares and passengers also enabled them to generate income for both the association and its

---

[8] For example, buses from Kariakoo to Manzese will share the large majority of their bus stops with buses from Posta to Ubungo, from Kariakoo to Ubungo, from Kariakoo to Kimara, and so on.

[9] The name 'railway line' was chosen because the queuing system implied that buses were operating in an orderly manner, one after the other, a bit like wagons on a train, rather than competing with each other, as was more common. It is important to note that workers from most routes that share only a few stops with other buses tend to adopt similar queuing strategies. Workers on routes whose stops mostly overlap with other routes cannot take advantage of such a measure. If they queued at the end of their route and workers from overlapping routes did not, they would lose custom. That is why *Mbele kwa mbele* ('Always ahead') is the most common practice among *daladalamen*.

individual members. Queuing of buses at the ends of the route should have made the task performed by *wapiga debe* redundant, as passengers could have easily found their own way to the first bus going to their destination. Yet the task was preserved as a pillar to establish both a rotating source of income for individual members of the association, as well as a collective welfare fund. Concerning the former, all association members took their turn, in pairs, to fill the buses. Given the number of members, these 'turns' took place approximately every two months. One's turn consisted of sitting on a bench at the station, noting down the plate number of each departing bus, from which they claimed the standard fee demanded by *wapiga debe* (150 shillings, or the cost of a bus ticket, worth approximately US$0.22 at the December 1998 exchange rate). Although both the number of buses in service and the number of rides per bus vary on a daily basis, the total amount collected by the association's pair of *wapiga debe* at the end of a day was significant, between 40,000 and 60,000 shillings (approximately US$58 and US$88 per pair, respectively, at the December 1998 exchange rate).[10] The association was then able to generate income for its welfare fund in two ways. First, every day each '*daladalaman* with a livelihood' paid the cost of one ticket into the association fund. Second, each day the association deducted a fixed commission from the two *wapiga debe* (a total of 10,000 shillings or US$14.70 per pair).

Workers 'with a livelihood' and *day waka* made up the vast majority of the association's constituency from the outset and throughout its existence. In the list of members for late 2001, these categories made up 110 out of 121 members (86.6 per cent): fifty-four (42.5 per cent) '*daladalamen* with a livelihood' (working on the twenty-seven buses serving the route) and fifty-six (44.1 per cent) *day waka*. Of the remaining eleven members (13.4 per cent), six (4.7 per cent) were *wapiga debe*. By allowing these men access to income-generating shifts and to the saving fund, the *daladalamen* pre-empted the retaliation that would have come from their being sidelined. The final five were 'absentee members'—former workers on the route who had found employment elsewhere, in most cases on *daladala* in other routes, but retained a bus-filling shift by continuing to contribute a fee to the association on a daily basis.[11]

The analysis so far has documented the contingent reasons that both prompted and allowed a group of workers to bring some regulation to the modalities of service provision and to set up a common fund. That its

---

[10] The variation depended on how many buses were operating on the route at a given time (from twenty to thirty), and on how many trips in a day each of them was able to complete (fifteen on an average day in 2002).

[11] There is no single file containing the total number of members in late 2001. The total has been worked out by consulting the names that appeared on the roster of bus-filling shifts for the last four months of 2001.

members could earn between US$21.65 and US$36.65 approximately every two months is some feat, the significance of which should not be underestimated nor overemphasized, given the relatively small sums workers earned through these shifts and the long intervals between them.[12] The way in which funds were spent reflects the values and goals that underpinned workers' collective action.

There are two ways in which members could draw on the funds of the association: as grants or loans. Grants were strictly limited to supporting members in three circumstances: the burial of people from their nuclear family; covering the health expenditure of hospitalized members; and bribing authorities to release members arrested for offences (real or presumed) against road safety rules. Loans were awarded on a case-by-case basis. Members' written requests to the association asked for support to cover, for example, hospital expenditure for their households or to purchase food when facing food shortages.

The association therefore provided its members and their families with an informal source of welfare protection and a social wage. Access to the association funds, while they existed, partly mitigated the severe state of insecurity and material deprivation in which unprotected workers lived. At the same time, to stress the political potential of this instance of associationism as being able to 'change the rules' of the game (Tripp 1997), or to claim that its 'generative powers' (Pieterse 2008) were capable of creating 'a very different kind of sustainable urban configuration' (Simone 2005: 4) would be highly misleading. The association's goal was to help members to manage the consequences of precarious employment rather than to challenge its causes. Furthermore, this workers' institution did not engage in the 'struggle between classes'; no claims were made on the state or on employers.[13] In addition, its money was systematically used for bribing policemen or jail personnel. Whilst essential to help members in trouble in the short term, in the long term it could be argued that this also helped to maintain the very socio-economic relations that generate the precariousness of the workforce. As the chapter has shown, it is the uneven and unregulated power relations between employers and workers that compel the workforce not to comply with road safety regulations in their struggle for survival, thus making them vulnerable to the police.[14]

[12] The two figures on individual earnings are calculated by subtracting the association's commission from the total sum that pairs could obtain from one day of 'work', and then dividing the result by two.

[13] The only exception is that the association reported the occasional invasion of its route by other unlicensed buses to the transport authorities. Such activity should be understood as a move to defend the lack of shared sections of that bus route, on which the very existence of the association depended.

[14] It is interesting to note that during fieldwork in 2009 several informants claimed that traffic police on the street were subject to severe pressure from superiors to extract bribes. They claimed

4.5.2 *The 'Struggle over Class'*

The tensions between individuals and different categories of transport workers in the daily life of the association further highlights the limits of often unwarranted claims about solidarity among the poor. The oversupply and fragmentation of labour visibly impacted on workers' collective identity, ultimately eroding their common ground. Tensions over the use of the welfare fund and entitlement to the income-generating shift of filling buses were the contingent elements of the 'struggle over class' amongst Dar es Salaam bus workers. Three kinds of tension emerged. First, from the outset there existed a structural imbalance between members' need for support and the association's capacity to provide it. One month after its foundation, on 20 December 1998, the chairman, opening a meeting to discuss 'the problems of shifts and how to help each other', stated:

> An outrageous series of problems has emerged. Problems that are so big that even the purpose of saving money has been lost. People do not understand why we save money. Consequently, everyone with a problem expects money to come from the association. If a licence [of a driver] has expired, someone wants money from the association; his dad is ill, he expects money from the association; he has not eaten yet, he wants money from the association...I remind you that money can be issued if someone has lost a close relative, is in hospital or is arrested by the police...But outside these cases, the money of the association cannot be touched.[15]

The struggle that developed over the right to bus-filling shifts reveals the second source of tension between transport workers, as different categories of workers, rather than individual workers, argued heatedly over the allocation of these shifts. Addressing the issue, the chairman put forward his own understanding of what constitutes a 'transport worker', and of the entitlements that came with it. Part of the audience disagreed:

CHAIRMAN: Let's move to the problem of shifts. Drivers and conductors all of us here. I don't think that there are other groups.

A: There is another group.

CHAIRMAN: Which group?

A: People who are jobless.

that in the same way in which a bus worker is expected to deliver a certain amount of money to the bus owner every day, a police officer stationed on the streets was expected to return a certain amount of money earned through bribes each week. Failing to do so would, it is claimed, result in transfer to a post which lack of contact with the public made less lucrative.

[15] Meeting of the transport workers' association, Dar es Salaam, 20 December 1998. The author attended the meeting, the audio recordings and transcripts of which are in his possession.

CHAIRMAN: How is it called?

A: People of the pavement (*kijiweni*).[16]

CHAIRMAN: Let's move to the problem of shifts. There are some of us, drivers and conductors who are really jobless, they sit on the bench and I would like to say that to do your shift [of filling buses] there at the station is not a job. It is a little extra for every man from Z. Whether you are a driver, a conductor, or jobless, a little sum of money earned quickly is useful for everyone. But do not consider it a job...

B:[17] Sorry, but it is absurd that you say that to fill a *daladala* is not a job when you have a job and I don't. For us it is essential. I can even go for two weeks without one shilling and suddenly I earn 10,000, 15,000, 20,000 shillings. A big help if you think that you get your 2,000 shillings every day. We would earn the same money so it doesn't make any sense when you say it is not a job.

CHAIRMAN: Listen B, many of us there at the station have our jobs. And if you have a job you cannot do another... If you want to tell me that this is a job you are wrong because a job is what you get up for every morning and you know that your job is waiting for you. That is what a job is.

C: If the work pays you, if it feeds you, and if you know that's where you find the money to eat, it is a job. And I, the one who has no *daladala* and sits on the bench, am telling you this...

CHAIRMAN: I repeat that filling a bus is not a job.

D: Again this story!

CHAIRMAN: As it is not a job everyone has the right to it.

D: As it is not a job, it is for those who are jobless.[18]

A third instance of tension among transport workers is the practice of the trading of the right to shifts to fill buses among association members, and the disagreement among members over the fairness of this practice. The association was set up to help a group of workers alleviate the harsh consequences of the commoditization of labour. However, the right to fill buses became a commodity in itself and the way in which it was bought and sold vividly shows, in line with Davis, that in the informal economy 'petty exploitation

---

[16] *Kijiwe*, literally meaning 'pavement' in Swahili slang, indicates the spot where the unemployed and underemployed congregate to look for work.

[17] Tragically, at the time of fieldwork in 2009, I learned that B was caught stealing from a house in Masaki and was stoned to death by the house's resident and its security guard.

[18] The way in which the association dealt with these antithetical positions among its members reflected a mix of compromise and of unilateral exclusion of some members by some others. The association formalized that a handful of those members who were jobless workers were accepted as beneficiaries of its bus-filling shifts and its informal welfare system. At the same time, it also decided that many other jobless workers were to be excluded from it.

(endlessly franchised) is its essence' (Davis 2006: 181). Those who sell their shifts, typically in advance, pocket less money than if they performed the task themselves. People sell out of sheer necessity, due to the need for money prior to the date of their scheduled shift, or fear of arrest (as the practice of filling buses is illegal).

The vice-secretary openly criticized such trade and the logic of petty exploitation that underpinned it, by warning its members that 'to sell one's shifts is equal to expelling oneself from the organization'.[19] Yet his warning carried little weight among association members as a vibrant trade in shifts was observable in 2001–2, and again in 2009. To give a sense of its significance, out of the forty-six shifts scheduled for the period from 31 August 2009 until 21 September 2009, twenty (i.e. nearly half) were sold.

4.6 Transport Workers' Horizontal Mobility and its Implications

The above instances of tension among transport workers, the dialectic confrontations over what constitutes a 'transport worker', and the dynamics of petty exploitation between them vividly illustrate the contingent ways in which transport workers in Dar es Salaam are fragmented and engaged in 'the struggle over class' (Harriss-White and Gooptu 2000). It is crucial to add some dynamism to the static picture of workforce fragmentation so far presented, by exploring patterns of labour fluidity and their political implications. Two key insights from research on the functioning of the informal labour market in India are a useful starting point. First is the existence of high levels of horizontal circulation of unskilled 'footloose labour' across different sites, as Breman (1996) eloquently described it. Second is the rigidity of the segmentation of the labour market, as ascriptive social institutions such as caste and religion make the transition from one subcategory of wage worker to another almost impossible. By contrast, ethnicity and religion do not regulate access to different categories of transport work in Dar es Salaam. The experience of being a transport worker revolves around great horizontal mobility across subcategories of transport work, from being a '*daladalamen* with a livelihood' to a worker 'on the bench'. But how quickly does labour turn over, and what are the patterns of labour fluidity in the sector?

In the absence of data on the workforce as a whole, the association's 'archive'[20] is an invaluable source in this respect as it constantly checks,

---

[19] This handwritten comment appears at the bottom of an untitled association file containing the roster of the shifts: a copy of the file is in the author's possession.

[20] The archive of the association consisted of two plastic bags containing the following types of documents: the list of buses and members of the association; the rosters with shifts at the station;

updates, and records the number of buses operating on the route, as well as the names and total number of members who were in the roster of bus fillers.[21] Analysis of these data suggests that workers changed status or moved elsewhere frequently.[22] As a result of the departure of some buses from service (due to breakdown, accident, or sale), the arrival of new buses, and changes in personnel (due to workers' illness or disagreement with owners), roughly one fifth of the 132 members had a new status in 2002, either as new 'workers with a livelihood' or as newly benched *day wakas*.[23]

The constant state of flux in which workers find themselves begs another question. Given the significance of access to this island of welfare for the workforce, and the accompanying oversupply of workers in the first place, understanding the dynamics of inclusion in, and exclusion from, the association becomes critical, as they ultimately determine the beneficiaries of its financial support. 'Urban space is never free', as Davis crucially points out. Indeed, access to the 'job' of filling buses in Dar es Salaam, similarly to 'a place on the pavement, the rental of a rickshaw, a day's labour on a construction site, or a domestic reference to a new employer...require(s) patronage or membership in some closed network' (Davis 2006: 185).

Indeed, there appears to have been a sort of 'carrying capacity' of the association, a threshold of the number of people it could support and beyond which it could not go. The high horizontal fluidity of its labour force went hand in hand with the fairly constant number of its total members: 121 in December 2001 and 132 in June 2002. However, the dynamics of inclusion in the association related to factors largely outside the control of the workers. New buses began operating on the route, and their '*daladalamen* with a livelihood' joined the association, as this brought advantages which were obviously too substantial to turn down. This constant influx of people was then offset by those leaving the association and moving elsewhere in their struggle for survival.

The dynamics of exclusion partly had to do with factors outside workers' control but partly with the fierce competition among workers for a foot in the door of the association. As the number of association members increased due to new arrivals, because workers 'with a livelihood' lost their buses due to their sale, due to 'misunderstandings' with the owner or a colleague, or to illness,

---

various letters of request for help; and a record of money. Copies of the files used for this chapter are in the author's possession.

[21] Knowing the number of buses in service and the number of members had important implications for planning the roster and for potential access to the association's funds.

[22] The frequency of labour mobility for the period December 2001–June 2002 was discerned by comparing the lists of buses and members serving the route in late 2001 with those for June 2002. The author generated the lists of buses and members for June 2002 by asking route workers what had changed from the 2001 lists.

[23] See Appendix B, which presents in detail the data to justify this step of the argument.

the number of bench workers also increased. Once 'benched', they joined those already waiting for a '*daladalaman* with a livelihood' to subcontract part of the working day to them. This created an oversupply of bench players in comparison to the number of jobs available to be subcontracted. As a result, a cruel scramble for scraps of work—another form of the 'struggle over class' (or indeed the struggle to be part of it)—took place among association members. The weakest members of the association simply did not obtain enough offers of work to get by, and moved on elsewhere: 15 per cent of the 121 members in December 2001 were no longer members in June 2002. But what constitutes weakness/strength among players on the bench? Times of service on the route, trust of colleagues, and family connections with 'workers with a livelihood' were significant predictors of receiving support through *day waka* when in need. As significant was the choice by 'workers with a livelihood' of whom they supported and how much they supported colleagues on the bench. Those who had not helped many *day wakas* when they had been '*daladalamen* with a livelihood' were rewarded with the same currency once on the bench.

4.7 United they Stood, Divided they Fell

This chapter has begun to explore the political sociology of transport workers. The analysis started by exploring the motivations behind the narrative that criminalizes the workforce as the main cause of transport problems. This necessitated investigation into workers' apparent lack of voice and political quiescence vis-à-vis bus owners and the state, a status quo which, at least in terms of public visibility, was maintained until 2002. Review of these workers' position in the broader economic structure showed that while they commanded low 'marketplace power', the specific economic subsector, urban bus public transport, in which they worked, conferred on them, at least potentially, 'workplace power'. However, factors such as the oversupply of workers, the fragmentation across different types of transport worker subcategories (with different roles and stakes at 'work'), and their spatial dispersion across thousands of atomized units of labour (buses), acted as barriers to the political organization of the workers.

The chapter has also explored the existing forms of solidarity among some of the workforce, and their significance, notwithstanding their failure to make demands on employers and the state. Through their initiative, workers generated collective savings which were spent to provide members with an informal source of social wage and welfare protection. Important as such efforts were in helping workers to manage the effects of precarious employment, they did not aim to challenge its causes. As such, the potential of these actions by 'ordinary' citizens should not be overstated nor romanticized. The association, as much an expression of solidarity as of chronic internal tensions, continued to exist up until 2005, six years after its establishment. The collective pooling of funds and informal welfare came to an end for reasons that were contested at the time of fieldwork in 2009. Its leaders blamed the turnover of the workforce and the unwillingness of former colleagues to contribute to the collective fund as the key cause of the demise of the association. Several members instead blamed the leaders for stealing the funds of the association. It is difficult to assess the merit of these two explanations, and the two accounts are compatible, as the high fluidity of its workforce might have triggered doubts about the sustainability of the association among its leaders, which ultimately resulted in their misappropriation of its funds. Furthermore, not everything was lost following the break-up of the association, as workers retained some degree of market self-regulation by preserving the queuing system at both ends of the route and the income-generating rotation of filling the buses. However, regular collections were no longer made, nor saved. Instead, ad hoc collections of money were and still are organized when 'people from the route' are in need. On balance, such developments made workers even more exposed to the whims of the labour market as, following the break-up of the association, their cushion to manage precariousness no longer existed. At the same time, their residual collective initiatives are evidence that some degree of workers' collective consciousness survived. How and when such consciousness was rekindled and deployed towards more ambitious political goals will be the focus of the next chapter.

5

The Politics of Labour 2

Struggling for Rights at Work (1997–2014)

5.1 Informalization and Rights at Work

Previous chapters have outlined the significance of class differentiation within the private minibus sector which provides public transport to Dar es Salaam and the harsh working conditions and employment insecurity faced by its informal workforce. The analysis has also investigated the factors that prevented, up to the second half of the 1990s, the political organization of the workers. A review of these workers' position in the broader economic structure emphasized that while they commanded low (labour) 'marketplace power', the specific economic subsector, urban bus public transport, in which they worked, conferred on them, at least potentially, 'workplace power'. As private buses have long constituted the only means of (barely affordable) motorized public transport available to the public in Dar es Salaam, unrest by its workforce has the potential to seriously affect the mobility of the vast majority of Dar es Salaam's population. I have also noted that workers' inability to take advantage of their 'structural power' was caused by their lack of 'associational power'. The activities of an informal association established by some of these workers, observed during fieldwork in 2001–2, helped workers to manage the effects of precarious employment rather than to challenge its causes.

Drawing on later rounds of fieldwork in 2009, 2011, and 2014, the focus now moves to understanding the causes and effects of workers' more assertive political attitude towards employers and the state. The chapter analyses the events that led to the foundation of the transport workers' association in Dar es Salaam—a process that first began in 1995 and that acquired public visibility in the early 2000s—the goals that it set itself, the strategy to achieve them, and the extent to which these succeeded. The significance of this close-up study of workers' mobilization is twofold. First, as one stop in the journey through the political economy of public transport in Dar es Salaam, its find-ings are to be read alongside and as a continuation of those presented in the previous chapter. They allow us to track the way in which, for workers in the same occupation and in the same place, the realms of possibility changed over time. Studying their political organization thus allows us to understand how workers went about overcoming their lack of 'associational power' and with what consequences on the balance of power between themselves and employ-ers. Second, stepping out of Dar es Salaam, the interest of this case study lies in the actors who are its protagonists, namely African workers in the informal economy in partnership with a trade union, and in the goals which workers' political mobilization can (or cannot) achieve in increasingly liberalized and informalized economies under neoliberalism. As such, this is a grounded contribution to the broader debate about the possibilities for collective action by informal workers today.

Leading contributions to such debate, not unlike much writing on the African city and on its informal economy, suffer from either a deterministic emphasis on the impossibility of struggles for rights at work, as structural changes in the way economic activity is organized undermine their potential, or from a search for human agency not matched by adequate attention to economic structures. Starting with the pessimists, the pervasive informaliza-tion of work since the 1970s has hit trade unionism hard. Trade unions have seen their membership shrink, and by and large have been unable to reach substantial numbers of informal workers, prompting a debate on whether globalization signals the end of trade unionism. As the majority of workers today operate in the informal economy, central to this debate is the role trade unions might play in organizing informal workers, quite apart from embra-cing broader goals.

A now widely held view argues that, due to increasingly informal employ-ment relationships that do not conform to any direct employer/employee relationship, workplace labourism is no longer viable. Take, for instance, Standing's (2011) highly influential voice on the study of labour under globalization. Standing is a leading advocate of the need to 're-embed the economy in society', arguing that although 'In principle trade unions could be reformed to represent precariat interest', in practice they cannot reasonably be expected to play any central role in this process. In his words (2011: 168), 'Trade unions lobby and struggle for more jobs and a larger share of output; they want the economic pie to be bigger. They are necessarily adversarial and economistic'. As 'who or what was the enemy' is no longer clear, trade unions are deemed to be institutions whose role has been made redundant by the informalization of employment.

Others, while less pessimistic about the relevance of unions to workers in the informal economy, share a negative view on the place that rights at work might play in their future agenda. Gallin (2001: 536–7), for example, suggests that 'only by organising the informal sector can the trade union movement maintain the critical mass in terms of membership and representativity it needs to be a credible social and political force'. However, as for the goals around which organizing will take place, Gallin agrees with Standing in suggesting that workplace claims and the formalization of precarious forms of employment cannot be the bread and butter of unions. The observation that the 'direct employee/employer relationship' has given way to 'more diffuse and indirect relationships'—including self-employment, paid work in informal enterprises, and casual work without fixed employers, often in combination—leads Gallin to conclude that 'trade union organising can no longer focus primarily on the employment relationship'. What is at stake instead is 'not formalising the informal but protecting the unprotected'.

Other scholars share this pessimism about the possibilities for struggles over rights at work without endorsing the primacy that broadening access to social protection must take. For example, Castells and Portes (1989) and Castells (1996) suggested that as a consequence of informalization, work heterogeneity caused the disappearance of the 'industrial and service proletariat class' (Castells and Portes 1989: 31–2), and that classes in the future 'may become more defined by their struggles than by their structure'. Davis argued that this pessimism 'went too far' as 'informal workers, in fact, tend to be massively crowded into a few major niches where effective organization and class consciousness might become possible if authentic labour rights and regulation existed' (Davis 2006: 185). However, despite their differences, both contributions are similarly reductionist. As this case study will aim to show, the existence of labour rights and regulation is the outcome of struggles by informal workers rather than a precondition for their existence or a simple function of unchallengeable structural changes in the economy.

Overall, these contributions successfully capture the current trends that have been disadvantageous to traditional labour organization in developed and developing countries alike, and ultimately to its crisis as 'traditional relations of representation and hegemony construction have been thrown in disarray and trade unions are no longer the undisputed articulators of mass discontent' (Munck 2013: 755). The way in which these trends have manifested themselves in individual countries is context-specific, due to uneven patterns of incorporation of countries in the global economy and due to balances of power between labour and employers which are both country- and sector-specific. But a general trend all scholars working on labour and trade unionism must reckon with is the growing elusiveness of clear wage relationships due to the mounting informalization of work. The study of labour is therefore increasingly becoming the study of informal labour.

However, the pervasiveness of unclear and informal employment relationships should not lead to the presumption that new forms of organization cannot emerge and cannot be successful at challenging the status quo at work (Piven and Cloward 2000). Such a dismissive stance sits at odds with, and fails to explain, the *ongoing* occurrence and diversity of labour unrest across the world. In this light, a broader goal of this chapter is to contribute to the debate on the possible goals for organized labour by analysing one instance of workers' mobilization, and its partial success, in pursuit of formalizing their employment relationship. In so doing, this close-up study of one group of workers aims to pay due attention to the enormous structural constraints against political organization, but without falling into determinism and thus losing sight of the agency which informal workers can exert.

An obvious starting point for any reflection on the possibilities for organized labour in the twenty-first century is seminal work by Silver, who has mapped labour unrest on a global scale from 1870 to 1996. Silver documented its *geographical* shift with 'the relocation of production *within* industries', and its shift across sectors *over time*, to argue that the impact of globalization and of its distinctive restructuring of production and of labour relations 'is less unidirectional than normally thought' (Silver 2003: 6). Along the same lines, Doogan (2009) wisely reminds us that 'capital needs labour'. Its capacity to move and relocate across borders is not homogenous across sectors and requires political support to free economic investment from national regulations. While labour pessimists have clearly understood the way in which the ease of mobility of capital has translated into new types of vulnerability for workers (such as the loss of jobs due to capital reinvestment in more labour-friendly countries and/or workers' reduced bargaining power due to capitalists' threats of relocating their investment), they have failed to appreciate how it has also generated new types of capital vulnerability to labour resistance.

Making sense of ongoing labour unrest, of why and where it takes place, and how it has changed over the years, requires a political economy approach to the labour relations that mediate workers' participation in the economy and to the balance of power between those who own capital and those who work for it, no matter how indirect the relationship between the two appears to be. Above all, as Breman cogently put it, what is lost with overgeneralizations on the impossibility of struggles for rights at work today, 'is any fine-grained analysis of the particular national economies, each with its own industrial and employment history, whose comparison might genuinely extend our understanding of precarization' (Breman 2013a: 135). In that spirit, this chapter is a fine-grained study of the context of the struggle by Dar es Salaam's bus transport workers.

Making sense of political action by actors in the informal economy also entails exploring the complexities of constructing common ground between trade unions and the 'informals'. This is the main strength of much recent work on the politics of organizing informal labour in Africa, the further general theme to which this case study contributes. The study of efforts to realize this common ground, both successful and unsuccessful, has stressed the importance of a contextualized understanding of practical activity to this end, focusing on the agendas of both informal workers and trade unions and their relationships with the state. It has cautioned against generalizing 'about the possibilities or impossibilities for alliances on the basis of structural differences or innate affinities' (Lindell 2010: 19–22).

Empirical studies of organizing efforts in the informal economy have been at their most insightful when attention to the political behaviour of 'informals' has been rooted in an understanding of the structural position occupied by them in a given economy (Meagher 2010b; Andrae and Beckmann 2010; Boampong 2010). Other studies on the politics of informality, in contrast, suffer from political economy blindness (Brown and Lyons 2010; Jason 2008; Jimu 2010). Fundamental questions such as how an economic activity is organized, who owns what in it, and its social relations of production, go either totally unaddressed or are skirted around. The detailed activities around which workers' organizing takes place often go begging. Take, for instance, Fischer, who engages with the theme of trade unions and informal workers in Tanzania. Her primary research exclusively focuses on interviews with ten leading unionists in Tanzania. As a result, Fischer has 'more data about respondents' attitudes towards the topic than about their concrete activities' (Fischer 2013: 140) and virtually no findings regarding the work of the unions and their results. By contrast, this chapter investigates the way in which a common ground between a group of informal workers and a trade union was built by firmly locating workers' political actions within their economic structures and against the challenges that these structures pose to workers' agency.

A variety of sources inform the analysis. The Communication and Transport Workers Union of Tanzania (COTWUT) kindly allowed me to consult its files on Dar es Salaam passenger transport workers. The correspondence between the transport workers' association (UWAMADAR) and COTWUT provides first-hand and fairly atypical insights into the process of how two distinct institutions went about 'building' a shared notion of the exploitation faced by Dar es Salaam's transport workers and a strategy to address it. The correspondence between UWAMADAR and COTWUT (the Coalition hereafter) and the bus owners' association provides glimpses into how the workers and bus owners related to one other over time. Newspaper articles on the subject were used as sources of further background information on the activities of the Coalition and its relationship with employers and the state. Interviews with workers, leaders from the transport trade union and the workers' association, employers, as well as relevant state officers, were carried out to probe and expand the key findings emerging from newspapers and archival files on transport workers in Dar es Salaam.[1]

The chapter proceeds as follows. The next section analyses the political organization by workers since 1995, its goals, and the strategy that workers developed in conjunction with the Tanzania transport workers' union. Particular attention is paid both to the process through which the Union and the association constructed a shared meaning of 'the *daladala* worker', and to the division of labour between the two parties. The chapter then goes on to document the Coalition's partial achievement of its main goal: the establishment of rights at work through the formalization of the employment relationship between bus owners and workers. It analyses how the Coalition changed its strategy by responding to the challenges raised in negotiating and implementing a workers' contract within a previously informal labour market. The conclusion summarizes the main arguments of the chapter and reflects on how they relate to the broad themes to which it aims to contribute.

5.2 From Political Quiescence to Political Organization: Early Days, 1995–2000

A handful of drivers and conductors, who would later become the first leaders of the *daladala* workers' association, first began to think of establishing an organization to defend their interests in 1995, but 'had no clue where to start'.[2] As such, the idea lay dormant until the summer of 1997, when a group of *daladala* workers organized a meeting to reinvigorate the plan. The forty-two drivers and conductors in attendance agreed to investigate the steps involved in establishing an association, and as a result found out that, according to Tanzanian law, trade unions rather than associations have the right to represent workers vis-à-vis employers or the government. In response to this, the workers' delegation visited the Dar es Salaam branch of COTWUT. Their visit ultimately resulted in the Union's Dar es Salaam Secretary agreeing to return the visit to a larger group of members of the would-be association.

Heavy on formal protocol, the first visit of the Union City Secretary, in July 1997, was nonetheless notable for the way in which both parties made substantial efforts, from the outset, to build common ground. Such efforts

---

[1] I shared a draft of the manuscript of this chapter with UWAMADAR's leaders, who provided useful comments and corrected some inaccuracies. I also planned to deliver a paper copy of this article to COTWUT Deputy Secretary, Mr Semvua, but was sad to learn about his passing away in 2014.

[2] The history of the relationship between UWAMADAR and COTWUT is recalled in 'UWAMADAR speech before COTWUT General Secretary', 9 November 2000. The author has translated the titles and contents of the Swahili documents quoted in this chapter.

initially centred on achieving a shared understanding of the occupational problems faced by *daladala* workers and on identifying a strategy to address them. The welcoming speech by one of the *daladala* workers, having emphatically stated his 'joy for meeting [the Union] today as we did not expect to have an institution that listens to our cries',[3] introduced the unionists to the reality of working on *daladala*. He did so by illustrating some of the main problems faced by its workers: the possibility of being killed by school pupils 'hungry for school' and retaliating against workers' refusal to ferry them at the discounted rate that the government had set but not funded; and the lack of sympathy for workers by the general public, whose expectation that ill people and pregnant women should travel for free overlooked the financial implications of this for the workforce. The speaker also emphasized how the lack of 'associational power' was at the root of workers' plight when he added that 'all these problems came from not having anyone to protect us and by not knowing where to take our complaints'. Now that the group had potentially found an institutional partner to voice workers' grievances, it put forward the key goals towards which they wanted to work:

1. To lead us to claim rights from rich people. [We want] employment like in other sectors. If a worker is released from work, his employer should look after him.
2. To oversee owners and protect us from them legally, so that the government can benefit from the existence of formal employment.

Employment rights were thus put forward by workers as a goal from the outset. The first meeting ended with the workers' request to meet the Union's General Secretary to take the agenda forward. The Union City Secretary, in forwarding their appeal to his superior, strongly endorsed it. As he put it, 'you will remember that for a long time we have been trying to find a way to get these workers involved in our union but it was very difficult to get them'.[4] *Daladala* workers in Dar es Salaam constituted a highly visible constituency, numbering between 20,000 and 30,000 potential members. The request for a partnership with their prospective association was thus met with interest by COTWUT.

Such interest in turn reflected the new landscape in which Tanzanian unions found themselves, following political liberalization in the mid-1990s. A key force in the anti-colonial struggle in the 1950s, the Tanzanian trade union movement became heavily controlled by the one-party state shortly

---

[3] 'Temeke, Tandika, Mbagala, Shule ya uhuru branch'. Handwritten speech signed by Mlawa and Kayombo (both UWAMADAR's leaders later on), 13 July 1997.
[4] From Dar es Salaam Zonal Secretary COTWUT to General Secretary COTWUT, 'The establishment of the *daladala* workers organization', 8 September 1997.

after independence (1961). In Shivji's words, the way in which TANU reorganized trade unions was 'contrary to virtually every principle of voluntary organization of workers or trade unions' (Shivji 1986: 233) and this fundamentally curtailed the autonomy of unions from the party and their capacity to represent workers' interests. The control of the ruling party over organized labour eased 'as a side effect of multiparty democracy' (Fisher 2011: 128). New legislation, first in 1998 and then in 2004, went some way towards cutting the umbilical cord between the ruling party and unions. Most notably, membership in unions became voluntary and union budgets were no longer part of the ruling party budget. Instead, they depended on their capacity to secure membership fees.

Such formal changes have resulted in a contradictory scenario. On the one hand, the detachment of unions from the ruling party was widely perceived as far from complete; on the other hand, there has been increased occurrence of strikes, negotiations, go-slows, and use of workers' votes as part of new 'repertoires' of unions in Tanzania following political liberalization (Fisher 2011: 141). Within this context, a new facet of trade union activities has been their increased attention to the 'informals', albeit with limited success in reaching them at national level and with important differences in the degree of interest in informal workers across unions. Amongst them, COTWUT appears to be at the forefront of the struggle to engage with informal workers. It has attempted to organize lorry and taxi workers, in addition to *daladala* workers, to whom the analysis now returns.

The meeting between the Union General Secretary and *daladala* workers marked the beginning of the partnership between COTWUT and the association of *daladala* workers. It was also another step forward in refining the strategy to fight the cause of the association's members. The two parties agreed that priority should initially be given to meeting the legal requirements for the workers' association to exist. This was no small task, especially as would-be political organizations seeking formal registration in Tanzania faced a Registrar of Societies that 'possesses excessive powers' (ICFTU, International Confederation of Free Trade Unions 2006: 3). Furthermore, *daladala* workers had no legal expertise. It thus fell on COTWUT to support the would-be association in navigating the Tanzanian legal system. Almost single-handedly, the Union drafted the constitution of the association so that it could comply with the regulations of the Registrar. This process took nearly three years. As the association's chairman gratefully recalled, 'the draft constitution was sent back with requests for revisions nine times. COTWUT did not lose hope and took care of these revisions'.[5] The association was formally registered on

---

[5] 'UWAMADAR speech'.

7 April 2000. According to its constitution, and reflecting their mutual interest in each other, UWAMADAR was formed as an association in itself but also a branch of COTWUT.

5.3 The Construction of a Shared Meaning of Exploitation

The registration of the workers' association laid the legal foundations on which the transport labour coalition rested. Shortly after registration, the two institutions intensified their efforts to further 'construct' a shared understanding of the Coalition's objectives, and of their respective roles in achieving them. The correspondence between the two, whose members were very different by education and working conditions, provides some insight into this process. For example, in June 2000 the Union's Dar es Salaam Secretary warned his General Secretary, ahead of his meeting with UWAMADAR members, that the people with whom he was going to meet 'are not used to leaders of the workforce [i.e. trade unions]'.[6] The Union was not familiar to *daladala-men* either, and needed educating about the reality of being a casual worker within the Dar es Salaam passenger transport system. A meeting, called for that purpose, still left the Union's General Secretary unclear about the working environment of *daladala* workers. He therefore asked UWAMADAR to put in writing 'the issues mentioned at the meeting'. Two weeks later, he received a letter from UWAMADAR's General Secretary, entitled 'The problems that drivers and conductors get at work'. The synopsis that the General Secretary gave is worth quoting extensively, for it allows an unusual glimpse into one instance of constructing a shared meaning of workers' exploitation.

> First of all, a bus driver in town wakes up as early as 3 a.m. to go and get the bus wherever it slept, as many buses sleep at the owners' place.[7] After this he will start work which will end between 10 and 12 p.m. Many things usually happen to him, such as to be attacked by thieves, and escaping that, there is no escaping from being stopped by Traffic Police at least three times a day. But the owner does not want to know all these things. He only cares that his daily sum (*hesabu*) does not decrease. There are days in which it rains a lot, and there is no business. But the owner does not want to know this; if he gets a flat tyre, the owner does not want to know the hours that he struggled to fix the tyre.

---

[6] From Dar es Salaam Zonal Secretary COTWUT to General Secretary COTWUT, 'Request from the association of drivers and conductors of urban buses (UWAMADAR) to meet with you', 7 June 2000.

[7] This is a literal translation which aims to reflect the broken Swahili in which this document was written.

The daily sum that owners demand, for example for a DCM bus,[8] is 45,000 to 40,000 shillings...

Think about a DCM bus operating from Gongo la Mboto to Kivukoni [one of the longest routes in Dar es Salaam]. It consumes 80 litres of diesel per day. At 506 shillings per litre this makes 40,800 shillings per day. Now driver and conductors, if they pay for breakfast, it is 1,000 shillings per day, if they eat lunch, it is 1,000 shillings. If you add this up you will see what is left for workers to divide from the day. [40,000 (owner) + 40,800 (diesel) + 2,000 (food) = 82,800] The money you can get from a DCM is between 80,000 and, for a very good day, 85,000. Will there be a shortage of mess on the streets if you bear in mind that these people have no salary?...

Imagine that at times owners even tell you that work uniforms, you need to buy them yourself...

Given these circumstances will the driver avoid creating a mess in the streets? Will he avoid refusing to ferry school pupils? Will he avoid shortening the route so that he can get many trips to earn enough money?[9]

The letter went on to list the costs incurred by bus owners, to conclude that:

Taking into account all the expenses, taxes, and maintenance, there is still a need for owners to establish a salary for workers, drivers and conductors...

It is hard to explain thing after thing but these are the conditions...

In light of the above we ask COTWUT to sympathize with the workers so that it can help us so that owners give out salaries for workers. I hope that you will appreciate the importance of the problem and work on it.

The reference in the letter to the struggle to 'explain thing after thing' emphasizes the centrality that getting to know each other played in the early days of the Coalition. Two further aspects—its wealth of detail on the economics of passenger transport in Dar es Salaam and the repeated (and rhetorical) question on workers' incapacity to avoid 'creating a mess'—provide clues as to the Coalition's division of labour and strategy to promote transport workers' rights. The Union's intention was to support the cause of *daladala* workers 'from above'. This entailed drawing on its technical expertise in labour law and on its political connections. The details of *daladala* operations and costs were presented in response to a precise request by the Union. It foresaw that their lobbying efforts with key state officers for employment contracts would be objected to on the grounds that the business of passenger transport was not profitable enough, and hence contracts were not affordable for employers. Thus, similar to other instances of organizing efforts by informal workers

---

[8] DCM is a minibus model produced by Toyota.
[9] Secretary UWAMADAR to General Secretary COTWUT, 'The problems that driver and conductors get at work', 5 July 2000.

(Narayan and Chikarmane 2013), information—such as details quantifying the economic reality faced by *daladalamen* and the uneven distribution of the wealth created in transport—was used to support the Coalition's demands for a fairer redistribution. As for the letter's reference to workers' economic compulsion to 'creat[e] a mess', this formed part of a broader strategy in response to the discourse criminalizing transport workers. It emphasized that the financial pressures faced by transport workers lay behind their 'misbehaviour', and subsequently argued that a less chaotic and more secure transport system necessitated a more secure and better remunerated workforce (see Barrett 2003 for a similar strategy in the context of South Africa). Thus, in exploiting the public nature of the service provided by transport workers, the strategy was to frame their interests as part of a wider societal 'common good'. Trade unions therefore do not necessarily frame their demand for rights at work in narrow and economistic terms. The nature of the economic sector in which their members operate clearly affects their discursive options.

In order to be politically credible, such lobbying 'from above' had to go hand in hand with the support of workers for the cause—and to COTWUT and UWAMADAR as the institutions promoting it—'from below'. As noted by Fisher (2011: 140) with reference to Tanzania, trade unionists face the 'fundamental question of how union power is backed up. You may have the authority to speak on behalf of your members or to represent labour matters in general, but is there a supporting majority behind you, the official asks[?]' (Fisher 2011: 140; see also Adu-Amankwah 1999). Reaching a critical mass of members amongst *daladala* workers was UWAMADAR's responsibility within the Coalition. It was a goal in and of itself but also an essential prerequisite for any Union lobbying 'from above'.

This left the ball in UWAMADAR's court. However, reaching *daladala* workers amounted to an enormous challenge, common to organizing efforts by 'vulnerable groups' (Lindell 2010: 9), and one that had to be delivered within serious financial constraints. '[There was] no money to promote the issue in newspapers, or for organizing attractive events. [The only way was] talking to drivers and conductors, one by one, "You have been doing this job for many years. Tomorrow, the day after tomorrow how will it look?"' (interview with Mlawa and Mnkeni, 2009).

So there was an element of sensitizing workers to the importance of employment contracts, and of trying to break the short-term time horizon of *daladala* workers' attitude to work that was both an effect and a cause of workers' occupational precariousness, as is often the case with the poor (Wood 2003). The albeit small financial support from the Union to hold events at which UWAMADAR could advertise its agenda is worth noting here, as it suggests that the Union was prepared to invest some of its funds to promote the organization of informal workers. This helped, in a small but significant way, to partly address UWAMADAR's lack of funds and the lack of visibility that came with it.[10]

Having gained legal status, and some resources to act, the focus turned to UWAMADAR's recruitment strategy. Ending the unregulated nature of the employment relationship in the sector, which from the outset was the ultimate goal of the organizing drive, could not be reasonably achieved in the short term. As with other instances of organizing vulnerable (women) workers in the informal economy, shorter-term 'forms of practical support which had more immediate and visible returns' (Kabeer et al. 2013) were often essential to attract and retain members who were daily pressed by their precariousness. However, this was not without risks as promising short-term support measures beyond one's capacity to deliver has the potential to undermine the credibility of any organization. Different institutions have responded in different ways to this dilemma. The choice of explicitly avoiding promises of financial and other benefits is not unheard of (Barrett 2003, on transport workers in South Africa). However, the choice by workers' organizations to provide services to their members is more common (Bonner and Spooner 2011; Von Holdt and Webster 2008; WEAZ 2006).

A letter entitled 'The way to run UWAMADAR', circulated in March 2001 to its prospective local leaders, documents the recruitment strategy adopted by this workers' association. UWAMADAR leaders did not shy away from ambitious promises. In return for a small fee, UWAMADAR pledged the following package to its members: support of a lawyer's services for work-related legal cases; payment for the renewal of driving licences and to cover the cost of members' funerals; as well as provision of health care expenditure for members and their families. Last, but not least, the letter stated that 'the employment issue is a very important one', and promised efforts to end the lack of regulation of employment in the industry.[11]

Another significant element of UWAMADAR's recruitment strategy was the decision to utilize transport workers themselves to promote the association at a street level. As such, leaders were identified at individual stations/routes, and educated about the association's broad mission and more discrete goals. It was then the branch leaders' task to recruit more members. Such a strategy provided workers with some leadership over the recruitment drive. Evidence suggests that this approach raises the chances of success in

---

[10] This can be discerned from a number of letters documenting the trade union's positive response to UWAMADAR's requests of financial support from the Union for events to be held.

[11] 'The way to run UWAMADAR', 10 March 2001. The same document outlined the financial plan to make UWAMADAR financially sustainable. This entailed the payment of a daily sum (2,000 shillings) from each branch and the payment of fees by individual members (2,000 shillings to join and 250 shillings monthly). In addition, it proposed proactively looking for sponsors—including UNDP, JICA, and the Nyerere Foundation, which had shown an interest in supporting the organization.

organizing informal-sector workers (Gallin 2001; Bonner and Spooner 2011; Barrett 2003).

The broad strategy adopted by UWAMADAR was not without risk, especially because the limited resources of the organization could not reasonably finance the provision of the range of promised services. Unsurprisingly, therefore, as I learned during fieldwork in 2009, there were people who considered UWA-MADAR as 'cunning thieves' or 'useless' (interview with Kizito, 2009). However, UWAMADAR's remarkable success in recruiting members suggests that its gamble paid off. In 2003, UWAMADAR had 5236 members, approximately 44 per cent of the total (estimated) workforce (UWAMADAR 2003: 23).[12] With the legitimacy 'from below' that such membership conferred on the Coalition, the time had come to begin lobbying for employment contracts for *daladala* workers.

5.4 Labour Rights through Collective Bargaining

As argued earlier, the unclear nature of the employer–employee relationship in the informal economy, and the frequent presence of intermediaries amongst them (albeit difficult to identify), is central to the argument that demand-making by organized workers around labourist goals is a thing of the past. As the employer–employee relationship is unclear, collective bargaining over working terms and conditions, the main weapon traditionally deployed by organized workers to confront employers, is no longer a realistic option (Gallin 2001; Standing 2011; Devenish and Skinner 2004). While this argument accurately describes the main challenge faced by workers, it fails to explain why such a challenge cannot be overcome in some instances, as well as why labour unrest around labourist goals still occurs.

In the case of passenger transport in Dar es Salaam, the workers' goal was to spell out the employment relationship between bus owners and workers. It has been suggested, in a study of the same sector in urban South Africa, that the existence of an organization of bus owners helps the process of collective bargaining with employers (Barrett 2003: x). The Coalition's primary counterpart was DARCOBOA, the association of bus owners/employers, and it was with them that the transport labour Coalition aimed to negotiate a collective agreement. As such, a way to force employers to the negotiating table had to be found.

To understand the strategy adopted by the Coalition, it is worth recalling the 'structural power' commanded by *daladala* workers in a city in which

[12] This percentage is based on UWAMADAR's estimate that there were 6000 private buses operating in Dar es Salaam at that time (UWAMADAR 2003).

privately owned buses constitute the only means of (barely affordable) motor-ized public transport available to the public. A strike would seriously affect the mobility of the vast majority of Dar es Salaam commuters, with immediate knock-on effects on virtually every economic activity in the city and beyond. At the same time, this 'structural power' had limits. The possibility of a strike was constrained in a context of oversupply of unskilled labourers (low 'marketplace power') since workers on strike without contracts could easily be victimized by employers and lose their jobs. Therefore, the Coalition had to rely on a less overt form of pressure on employers in order to encourage state involvement and mediation between the two parties, just as has proven crucial in other contexts (Barrett 2003; Von Holdt and Webster 2008; Kabeer et al. 2013). Time and again, rumours occasionally reported in the press (*Nipashe*, 9 June 2008; *The Citizen*, 7 December 2009; *Tanzania Daima*, 29 March 2010; *HabariLeo*, 6 April 2011) would spread about a forthcoming strike by *daladala* workers. The Coalition would promptly deny any involvement with it—or be unavailable for comment—and yet on the day of the strike buses on *some* routes, or some buses on several routes, would be withdrawn from transport service provision for part of the day, causing disruption to passengers who would in turn complain to public authorities. Workers also resorted to violence, using stones to attack the buses of those workers who did not adhere to the protest (*The Citizen*, 10 December 2009). In the words of COTWUT's General Secretary, 'workers were stopping work when they wanted to complain about something. And the government mediated, called a meeting with DARCOBOA, UWAMADAR, COTWUT, to solve the problem' (interview with Semvua, 2011).

As Silver has pointed out, there is much to be understood by studying 'anonymous or hidden forms of struggle such as undeclared slowdowns, absenteeism, and sabotage [which] are especially significant in situations where strikes are illegal and open confrontation difficult or impossible' (Silver 2003: 35). In this instance, wildcat strikes and localized walk-outs were an effective strategy in that they exploited the structural power com-manded by workers. On the one hand, they were insufficiently confronta-tional to trigger widespread retaliation by employers; on the other hand, they were assertive enough to establish the demands of workers on the political landscape and to attract the attention of Dar es Salaam transport policy-makers. The latter, like the Tanzanian leadership itself, were facing growing tensions over the costs and impact of economic deregulation, to which they needed to respond. Public officials' desire to deal with transport workers' unrest with all haste led to their gentle but firm pressure on DARCOBOA to negotiate a solution to their grievances with the Coalition.

Forced to sit at the negotiating table, the *daladala* owners' association reacted ambiguously to the issue of contracts. As its chairman wrote to

UWAMADAR, 'DARCOBOA has no employees. Drivers and conductors are employed by private owners of individual buses.' Thus, the guidance that the association prepared was to be seen as no more than 'advice to owners, who will decide themselves, not DARCOBOA'.[13] At the same time, DARCOBOA also played an active role in the process of collective bargaining, and succeeded in including in the contracts elements that rendered their adoption difficult (see section 5.5). Furthermore, on one occasion, DARCOBOA's chairman even falsely claimed that owners were 'the ones who proposed this [i.e. employment contracts] to the government as a way of reducing accidents' (*The Guardian*, 5 August 2009).[14]

For all the bus owners' delaying tactics, the negotiation over workers' employment contracts proceeded slowly but surely. A collective agreement became legally binding on 26 March 2004, following the seal of approval by the Tanzania Labour Court. As Mr Semvua, the Union Deputy General Secretary, recalls, the collective agreement 'had three things, big ones. It spelled out the employer/employee relationship which was not there until then. That contract mentions that the driver and conductors of a certain bus are such and such and makes them employees. It established a wage level and the working hours per day. It established the right to holiday for employees' (interview with Semvua, 2009). The contract legally brought to an end the unregulated nature of the employment relationship in the Dar es Salaam passenger transport sector that had been central to the bus owners' ability to financially squeeze workers. Seven years after the first meeting between *daladala* workers and the Union, the Coalition had therefore scored a significant achievement in advancing the cause for which it was set up.

5.5 Barriers to the Enforcement of Employment Contracts

As is often the case, *daladala* workers' entitlement to labour rights *de jure* did not translate smoothly into their enforcement. The reasons why the vast majority of workers did not enjoy employment contracts de facto were hotly contested by the political actors involved in bringing them about. The relationship between UWAMADAR and the Union turned tense over the matter, with the City Secretary of the Union blaming UWAMADAR for its failure 'to organize its members'. As he put it, 'The power of the Union is in the hands of

---

[13] Chairman DARCOBOA to Jimmy Mnkeni, UWAMADAR: 'The contract for decent work', 31 March 2003.

[14] General Secretary DARCOBOA to Zonal Secretary COTWUT, 'Seminar of *daladala* owners', 9 February 2004. Both UWAMADAR leaders and public transport officers suggested that such claims were false and had to be interpreted as an attempt by employers to downplay the strength of the labour coalition. See interview with Mlawa and Mnkeni, 2009, and interview with Sulemani, 2009.

its members. It is now up to the members to organize and start demanding the contracts.' He further added that the bus owners 'have deliberately been neglecting the legal contract, knowing that the drivers are not sufficiently organized to take action'.[15]

UWAMADAR, on the other hand, stressed that the characteristics of the workforce and of the labour market were a major obstacle to the mobilization of *daladala* workers to claim their right to contracts. First, from the initiative of the association of bus owners, the collective agreement established specific skills as a prerequisite for drivers and conductors to qualify for a contract. These included, for drivers, possession of class C driving licences. Given the low level of education of the vast majority of *daladala* workers, and that many amongst the workforce held class B licences, such conditions proved to be a spanner in the works for the Coalition (*HabariLeo*, 30 June 2007). As if this was not enough, the extremely rapid turnover of labour, a structural characteristic of the *daladala* labour market, also negatively impacted on the workers' association's capacity to reach out to members. As its chairman put it, 'You find a good branch leader, but before you know it work has taken him to another route. End of story' (interview with Mlawa and Mnkeni, 2009).

While the lack of education of the workforce and its occupational fluidity were the proximate causes of the slow enforcement of contracts, above all, as UWAMADAR reflected, 'the problem with contracts is that the government did not get involved'. While the role of the government in steering the process of collective bargaining between employers and workers had been fundamental, the contract had no built-in mechanism to ensure that the state oversaw its enforcement. From the ashes of their bitter exchanges in the press, the Union and UWAMADAR found new common ground in concluding that the widespread adoption of contracts required stronger support from the state.

5.6 Labour Rights: Bringing the State Back In

This is not to say that the Coalition had overlooked the central role to be played by the state. Given DARCOBOA's uncooperative attitude throughout negotiations, only three days after the collective agreement had been approved, the Union General Secretary, in writing to the Chairman of the Dar es Salaam Transport Licensing Authority (DRTLA), made it clear that the DRTLA [was] relied upon 'as a very important connection' in making this happen.[16]

---

[15] 'Contracts yet to materialise', *The Express*, unknown day and month in 2004.
[16] Dar es Salaam Secretary COTWUT to Chairman DRTLA, 29 March 2004.

However, once a collective agreement on contracts was legally signed, a more focused effort at strengthening its relations with the state apparatus became the main activity of the Coalition. The choice of who was invited as 'Guest of Honour' to the UWAMADAR annual meeting is a symbol of this shift in strategy. Until 2004, this role had been given to the General Secretary of Union, reflecting the primarily inward-looking focus by UWAMADAR on consolidating its alliance with its trade union partner. In 2004, for the first time, and then subsequently, the role was played by state officers. The first 'outsider' to be invited as guest of honour to celebrate the anniversary of UWAMADAR's foundation was Lieutenant Makamba, the Dar es Salaam Regional Commissioner at that time.[17] Such moves were not just symbolic. It was to Makamba that the Union leader wrote four months later, to complain that DARCOBOA did not attempt to influence their members to issue contracts for their workers, and that even DARCOBOA leaders did not issue contracts to the workforce of their own buses. The Union City Secretary also informed the Regional Commissioner that a strike was becoming unavoidable. He therefore called on the local government, 'the one who steers us all in this region',[18] to organize a meeting with DARCOBOA, UWAMADAR, and the Transport Licensing Authority to discuss the issue further. Guests of honour were therefore also key ports of call for the Coalition and, as such, their selection was strategic.

With time, the Coalition broadened its initial goals. It did so by forging alliances with other groups with whom it shared strategic interests. Three instances of the increased ambitions of organized workers illustrate this. First, joining hands with the Tanzania Drivers Association (TDA), the Coalition held meetings to publicize the 'problems faced by *daladala* drivers'.[19] A new goal was to challenge the ambiguity of the law and its negative implications for the workforce (see also Barrett 2003, on a similar effort by South African transport workers). The party took issue with drivers being held exclusively responsible for violations of road safety rules, arguing that there were 'violations which are obviously the responsibility of the owner. For example, the poor condition of a vehicle; not owning a transport licence; not having insurance; not having a vehicle inspection report.' It asked that 'the Road Traffic Act be modified to openly distinguish the violations for which owners or drivers are responsible'. Second, while the pressure on the issue of

---

[17] General Secretary UWAMADAR to Regional Commissioner Dar es Salaam, 19 April 2004.

[18] Dar es Salaam Secretary COTWUT to Regional Commissioner Dar es Salaam, 'Complaints against DAR(CO)BOA on the implementation of decent work contracts for drivers and conductors of *daladala*', around 29 July 2004.

[19] 'Minutes of a meeting on the conditions and problems of drivers with the Permanent Secretary Ministry of Infrastructural Development called by the Tanzania Drivers Association', no earlier than 13 July 2006.

employment contracts continued, it broadened to include more ambitious lobbying for changes to the regulation of the private sector, which would ease the enforcement of workers' employment rights. The Coalition urged the government to ensure that passenger transport was 'provided by companies instead of individual owners. This will make it easier to adopt contracts.' Furthermore, in response to the adoption of doctored contracts—containing the photo or details of a person who was not the actual driver, and prepared by some employers to avoid the regulation of labour relations—it called the Ministry of Work, Employment, and Development to prepare a blueprint of an employment contract. Third, it pushed for a stronger intermediary role by the state, and demanded that the newly established Sea and Maritime Transport Regulating Authority (SUMATRA) make the submission of a copy of workers' contracts a condition of issuing bus owners with a passenger transport licence (*HabariLeo*, 30 June 2007).

The attitude of public authorities to these demands was rarely one of cooperation at the outset. For instance, the state's initial response to workers' requests to address the ambiguity of the law amounted to a firm rebuttal couched in techno-legalist terms. The Permanent Secretary claimed that 'although these faults apparently are the sin of the owner, the driver is the temporary owner when he operates a vehicle. Furthermore, the law prevents the driver from operating a vehicle when it is not roadworthy or without important documents'.[20] Such words suggest that key state officers were not prepared to concede that the 'temporary ownership' of vehicles by drivers actually reflected the employers' strategy to transfer the uncertainty of returns from work onto the workforce, as we have seen. Yet, the concession by the Permanent Secretary that 'fear of losing their job' forced workers 'to drive the vehicle according to the preferences of the owner and against the law', shows some awareness of the uneven balance of power between workers and bus owners. Such an ambivalent stance, however, did not translate into any commitment by the state to meet workers' demands.

Having said this, the state's position softened over time, due to continual pressure by workers.[21] In 2009, the workers' call for a stronger intermediary role for the state in enforcing contracts was finally met when the registering of workers' contracts, inclusive of photos and signatures, became one of the requirements for the issuing of public transport licences. This was no small victory. As the Union General Secretary put it, 'at least workers now had a

---

[20] 'Minutes of a meeting on the conditions and problems of drivers with the Permanent Secretary Ministry of Infrastructural Development called by the Tanzania Drivers Association', no earlier than 13 July 2006.

[21] It is fascinating to note the way in which these negotiations disappear from the radar of newspapers and from the Coalition's correspondence, only to re-emerge years later.

place to start. Unlike the first collective agreement between owners and workers, the issue of contracts is now a rule of SUMATRA, a government office, under the Ministry of Transport' (interview with Semvua, 2009). Similarly, it was in response to the threat of another strike, in 2010, that the Minister of Transport declared that the traffic law violations of drivers, conductors, and of vehicles (i.e. their unroadworthy condition) would each carry 'their own weight' (*Dar Leo*, 3 December 2010). There was now a discernible move towards making the law less disadvantageous to workers.

Reflecting on the challenges which lay ahead, leaders of UWAMADAR and of the Union were under no illusion that making the enforcement of contracts a reality would be a straightforward process. As the COTWUT Deputy General Secretary put it, 'the biggest challenge in implementing the rule is the fact that *daladala* owners are accustomed to exploiting workers without contracts, so they will try everything they can to avoid this change.' However, while taking the likely resistance by employers into account, the Union leader emphasized the significance of the albeit marginal gains made by the Coalition. As Semvua put it, bus owners' room for manoeuvre in avoiding labour regulations was progressively shrinking: 'the day that an owner gets into an argument with his driver, and is asked to produce the contract, he will be in trouble' (interview with Semvua, 2011).

5.7 A New Political Subject: Trade Unions, the Informal Economy, and Labour Rights

Difficulties in holding employers to account, however, resulted in renewed and unsolvable tensions between the workers' association and the trade union, eventuating the demise of the Coalition. Leaders of the informal workers association grew disillusioned about the necessity of their partnership with the trade union, as they doubted its effectiveness in the continued struggle for labour rights. As the UWAMADAR General Secretary recalled, 'the service that we were getting was small, and our needs to be looked after were not satisfied' (interview with Mlawa, 2014). Frustrated by the lack of influence of UWAMADAR within the trade union, its leaders, together with those of the association of upcountry bus workers (UWAMATA), left COTWUT and began work to found the Tanzania Road Transport Workers Union (TARWOTU), which was officially registered in January 2013.[22] The COTWUT General Secretary was very dismissive about both the claim that his union neglected the interests of *daladala* workers and the motives behind the UWAMADAR leaders' decision to leave his union: 'the ambition to lead sometimes

---

[22] Lorry workers are the other source of members for TARWOTU.

drives change, the desire to be the General Secretary of a national union. Otherwise, why not use a network that is already in place?' (interview with Semvua, 2011). It is obviously difficult to ascertain which of these two competing versions of events better explains the split. Arguably, this does not matter, because 'If there is one thing that history has taught us, it is that trade union structures almost never develop smoothly by means of piecemeal engineering. They are generally the outcome of conflicts and risky experiments. Pressures from below (articulated in competitive networks, alternative action models, etc.) will be of the utmost importance in deciding that outcome' (Breman and van der Linden 2014: 937). The events analysed in this chapter, such as the emergence of a new trade union from the ashes of the Coalition, fully controlled by informal workers, neatly fits the dynamics of conflict and risky experiment that trade union renewal entails.

5.8 Contextualizing Workers' Power and Realms of Possibility

This chapter has analysed the political organization adopted by *daladala* workers since the late 1990s, their alliance with the transport trade union, and how this evolved over time. It has outlined the goals that the Coalition set for itself, the strategy it chose to achieve these, and the slow but incremental advances taken towards them over the period 1997–2010. The events analysed are context-specific. They are also open-ended as they reflect the outcome of a political battle, between different groups with conflicting interests, over labour rights for Dar es Salaam transport workers. While these characteristics do not make the case study replicable, a number of general considerations can be derived around the actors involved (African workers in the informal economy in partnership with a trade union) and the goals which workers' political mobilization can (or cannot) achieve in increasingly liberalized and informalized economies.

What can be learned about the relationship between trade unions and workers in the informal economy from this instance of a partially successful partnership between the two? The first insight is about process: realizing 'associational power' entailed a slow but sustained effort at constructing a common ground between informal workers and their Union counterpart. The complex nature of this process partly explains the very slow pace at which change took place and underlines the need to allow adequate timeframes when studying the politics of informality. The analysis has also shown the importance of informal workers' leadership to the success of their political organization. While the Union provided its political connections and know-how to support lobbying on behalf of transport workers, workers themselves were not only the initial trigger behind the formation of the Coalition, but also its key outreach workers, and the leading partners when setting the Coalition's goals.

Furthermore, the Transport Union's attitude towards minibus workers and their association sits uneasily with the characterization of unions' attitudes towards organizing informal actors as ineluctably uninterested or opportunistic. In this case, notwithstanding some tension between the two organizations and the split between the two in 2013, the Union was prepared to invest resources and energy in supporting the organization of precarious workers, the success of which was of significance to both the Union and workers themselves. Out of this initial synergy a new trade union has been born, fully controlled by informal workers. The contrast between this relatively positive story and less positive instances suggests the need to move away from overgeneralizations, towards nuanced and contextualized approaches to the study of trade unions and their relationship to economic informality and precarious workers within it. Indeed 'organised labour [can be] part of the solution as well as being a problem at times' (Munck 2013: 760).

The second general insight that can be derived from this case study on the politics of organizing in the informal economy is methodological, as the study shows the benefits of using class-based political economy as an analytical approach. Far too often, studies on economic informality skirt around the questions of who owns what, and with what outcomes, in the informal economy. By contrast, this study demonstrates that understanding the way in which the workers are linked to (capitalist) employers, locating workers within their economic contexts, and mapping the sources of both their precariousness and power is essential in making sense of why and how workers mobilize politically. Furthermore, functionalism and reductionism, features of some work in the political economy tradition, can be avoided. This study has not simply read off workers' political interests from their economic position in society, nor has it identified the economic position of workers as the only predictor of what was politically possible for them. Indeed, the very barriers that prevented the organization of these workers up to 1995 were then overcome through the initiative of some workers and the events that unfolded subsequently.

The final lesson to be learned from this study is that its findings sit uneasily with, and raise questions about, the widely held belief that collective action by organized (or organizing) labour along labourist goals belongs to the past, and that social protection is a more realistic and strategic target in tackling workers' precariousness.[23] The workers on whom this chapter has focused shared

---

[23] Doubts have been raised on the strategic superiority of a focus on social protection over and above rights at work as a measure to tackle precarity. As the political momentum behind universal social protection is non-existent in many developing countries, calls for its introduction lack the necessary pressure that is likely to result in its adoption (Lerche 2012).

the lack of a clear employment relationship to their employers—the main characteristic that, it is argued, prevents workers' mobilization for a 'rights at work' agenda in the informal economy. While this was indeed the main source of workers' precariousness at work, it was also the very stimulus and goal of their mobilization. Drawing on the research of Wright and Silver, this study has argued that Dar es Salaam minibus workers, when challenging the unclear nature of employment relations in the sector, drew on the significant 'structural power' they commanded as providers of the cheapest form of available public transport. While the circumstances and context in which these workers' mobilization took place are necessarily specific, that such workers could command a degree of structural power stresses the value of disaggregating the realms of the possible for different groups of workers in different economic sectors and countries. Furthermore, above all, it underlines the importance of putting ongoing labour struggles at the centre of reflection on the possibilities for action by precarious workers.

6

Tracing Occupational Mobility/Immobility among Informal Transport Workers

6.1 Hitting a Moving Target: Methodological Issues

What is the long-term impact of employment on *daladala*? What trajectories can one discern about the occupational mobility of its workforce over time? What lessons are there to be learned about upward mobility or immobility in the informal economy? Answering these questions is another important dimension in the search for human agency, often lacking attention in apocalyptic narratives on the African city. This chapter will set such agency against the backdrop of structural constraints with which it interacts and which are themselves seldom referred to in much of the 'rosy' writing on urban Africa.

Daladala workers engage with such questions through their own writing on buses, and these missives are therefore a useful starting point from which to begin thinking about the issues. For example, through the statements 'Life is round' and 'Money is round', two workers emphasize the unpredictability of life and of earnings. The possibility of making money is proudly boasted about by workers who write *Onja mafanyikio* ('Taste achievement') and 'Hard work pays'. In a similar vein, workers who write *Pesa Mbongo* ('Money if you are smart') and *Kuti Kavu kuanguka sio ajabu Trans*[1] ('It is no surprise when the empty coconut falls down the tree, i.e. it is no surprise that the empty-headed person does not make it') add to this a self-congratulatory ring, as they emphasize that cleverness or stupidity is what drives success or failure. A much gloomier message is put forward through statements such as 'So many tears', 'So many rivers to cross', 'The hard time', and 'Tough life'. They emphasize instead the difficulty of work. Other workers underline the centrality of this work to their livelihoods: 'Daily bread' and *Kitunze kidumu* ('Preserve

---

[1] 'Trans' is often used in writings on *daladala,* buses, and lorries as the abbreviation of transport.

it so that it lasts'). The meagre nature of earnings from work and the mismatch between effort and rewards is another trope, evident in writings such as 'Money torture', *Kiasi cha mboga* ('Enough to buy greens', i.e. but not enough to buy protein), *Posho nauli, kesho wahi* ('Your daily return today is worth a bus fare, tomorrow come earlier'), and *Posho musiki, kesho wahi* ('Your daily return today is the music [that you listened to in the bus], tomorrow come earlier'). One worker is more cynical still about the prospect of making money honestly, writing *Penye pesa hapakosi majungu* ('Where there is money there is no lack of dodgy deals'). Some workers seem to take the high ground, stating 'Whatever the case' and *Yote maisha tu* ('It is life anyway'). What most workers agree upon, however, is the need for a *daladala* worker to be tough, for a number of writings on buses are about machismo and power, in different forms. One sees 'Tuff boy' (sic), 'Bull Fighter', 'Fighter', 'Iron Man', 'Original gangsters', 'Power Viagra', and a parade of world leaders who are considered symbols of power, such as 'Polpoti' (Pol Pot), 'Arafati' (Arafat), 'Netanyahu', 'Osama wanted', the Nigerians 'Sani Abacha' and 'Mashood Abiola'. Notoriously tough areas, wars, and military weapons are another popular choice, seen on buses labelled 'Soweto', 'Baghdad', 'Gulf War', 'Desert Storm', 'Scud', 'Patriot'. Confronting such machismo, one worker comments *Nguvu ya mamba, nje ya maji* ('The strength of a crocodile, out of water').[2] Finally, illustrating the existence of different opinions on the issue among workers, while one suggests that there are rewards to be earned through the hardships, by saying *Chungu lakini dawa* ('Bitter but at least medicine'), another worker cautiously warns *Maomivu yakizidi mwone daktari* ('If the pain increases, see a doctor').

These are insightful comments, and attempting to achieve some quantitative understanding of the prevalence of the different scenarios at which they hint, and some sense of the way in which employment on *daladala* impacts on its workers over time, albeit in different ways for different groups of workers, felt important. Such an analytical agenda, however, presents serious methodological challenges. The informal nature of employment in the sector implies that there is no existing list of workers from which to start to build a sample. Furthermore, the nature of employment causes a very rapid turnover of workers, as we have seen in Chapter 4. In light of these circumstances, the research on which this book is based drew on a sample which was built by using the roster of a *daladala* workers' association as a starting point.[3] This recorded the names of all members entitled to shifts to fill buses, and thus to earn money from them. The latter characteristic gives credibility to the list of

---

[2] The strength of crocodiles is almost uncontrollable in water but it is drastically reduced once out of it. The metaphor captures the contrast between the authority and power that *daladala* workers display on the buses, and their weakness once out of them.

[3] The activities and significance of the association were analysed in section 4.5.

names that the roster contains, for no member of the association ever missed an income-generating shift. After all, *Maji mengi, unga kidogo* ('Lots of water, little flour'), as one *daladala* driver reminds us, was their experience.

What was invaluable about this roster is that it allowed identification, out of the seemingly endless flux of workers that one could observe, of a number of people who worked on a particular *daladala* route in 2001. In a sense, it acted as a class photo. Drawing on the registers of shifts over four months, a list of 121 workers was generated.[4] Against this list, I recorded their occupation twice, in 2009 and in 2014. In the first instance, this was done with the help of three workers—whose names also appeared on the list—who acted as informants.

Two of these informants provided information as a pair whilst one worked on his own. These particular workers were selected for two reasons: they were all 'on the bench' at the time, and they had all worked the route for over twenty years. As such, they had both time to go through the list and familiarity with the names of workers appearing on it. As anyone with experience of fieldwork can easily imagine, the exercise did not often proceed as schematically and tidily as suggested above. Other workers would help out, confirming and/or suggesting more up-to-date information on the whereabouts of individuals on the list. Also, this parallel process of tracking the occupation of workers did not always produce a consistent picture, as the two sets of informants suggested different occupational destinations for nine of the 121 workers. While the limited number of cases on which there was disagreement is reassuring, and revealing about the long-lasting nature of some degree of connection between these workers, choices had to be made on how to deal with these nine cases. In the first instance, I notified informants about the different occupations suggested by their colleagues. In five of these nine cases it appeared that one of the informants was more up to date with the destination of their colleague, who had recently moved to another job. In four cases, however, I was not able to reconcile the inconsistent suggestions. I therefore followed the information provided by the two sets of informants for two cases each. The low number of cases, four out of 121, does not affect the overall picture in a dramatic way. A further step in trimming the occupational list was that the occupational whereabouts of three of the 121 workers was unknown to my key informants, while two names from the roster did not resonate with them. The total number of traceable workers was therefore 116. Finally, eleven workers had died between 2001 and 2014, three of them before the 2009

---

[4] While 124 names actually appeared on the register over six months, three workers were double-counted as their nicknames, alongside their official name, were used on the rosters. There was no consensus about the occupational whereabouts of three of the workers, as further explained in this section.

research spell. The total number of active workers for which information could be gathered therefore changed from 2009 to 2014. In 2009, the total was 113; in 2014, it fell to 105.

The aspirations of *daladala* workers for their own progression from employment on urban buses is best captured through the words of Dongo, a *daladala* veteran. As he explained, workers hope that employment on *daladala* is to act as a 'passing-time job'. As Dongo further explained:

> A big goal when you work on *daladala* is not to stick here, as if you made it. That is why you find that some people have moved on to other jobs. You work on *daladala* because you have no other job. To get enough to get by. If the opportunity to get a proper job comes up, with a wage at the end of the month [you would grasp at it]. (Interview with Dongo, 2014)

As his thoughts were shared by all workers with whom I discussed occupational mobility, this longitudinal tracking exercise aimed to track how many workers were able to fulfil this goal. This meant that there were four trajectories of employment mobility/immobility, corresponding to four groups of workers:

1. *daladalamen* 'forever' group, e.g. those who still worked on *daladala* in 2014;
2. *daladalamen* 'no more', e.g. those who by 2009 had changed occupation and who had not returned to work on *daladala* in 2014;
3. 'out and back in' *daladalamen*, e.g. those who by 2009 had changed occupation but had returned to work on *daladala* by 2014;
4. 'slowly out' *daladalamen*, e.g. those who were working on *daladala* in 2009 but were found to have changed occupation in 2014.

Following analysis of the list, I conducted semi-structured interviews with twenty-five of these workers. Their focus was on workers' own individual occupational trajectories, on what workers achieved out of their work on *daladala*, but also on how their work on *daladala* compared to other occupations. Such discussions with individuals about their own individual trajectories often trespassed onto broader considerations about what the experience of being a *daladala* worker means in general. Interviewees often compared their own trajectory to that of former colleagues, in so doing putting forward their own understanding of agency, and of how individuals experienced and navigated in different ways the same structural pressures.

The selection of the twenty-five workers that I interviewed for the longitudinal component of this study was dictated by both analytical and logistical considerations. On the former, this component of the study did not take place in an analytical vacuum. Interviews with workers highlighted the importance of paying attention to the differences between workers if any understanding of who *daladala* workers are was to be achieved. In the words of former *daladalaman* Tolu:

> What type of driver? You must look at him from close and assess. Who do you live with? One lives at home with his parents. So he is only looking for money to eat. But the one who is looking for development, and has not started yet, and starts off as a conductor, he stands no chance. The man with a family has this pain: eh bwana, if I don't make any money today at home what will they eat? A man with no family, what he is looking for is 2 or 3,000 shillings, he will eat chips in poor areas (*uswahilini*), and he will be on the road the day after. He is also not working in the same way. You wake up, maybe at 5 a.m., but you are not young, so at 8 p.m. you need to stop. The beauty of being young is that work never ends. If there are no more passengers in Mwenge, you go to Tegeta. From Tegeta, to another area. You have no family, nor a house. The girlfriend, you have sex with her the day in which the bus is serviced (sic). Those with children cannot work like this.
>
> (Interview with Tolu, 2010)

The choice of workers whom I aimed to interview thus reflected the variety of occupational trajectories and personal circumstances: workers of different ages, workers who had not moved on from *daladala* and workers who had, and workers with a variety of post-*daladala* occupations. Logistics also mattered, as the location of these workers (whether in Dar es Salaam or not, and where specifically in the city), their availability of time for an interview in their new job, and access to their mobile phone number to arrange an interview, were important factors in shaping up the final list of actual interviewees. Before presenting the substantive findings of this research effort, a note of caution is due on the limitations of these findings, which, it is important to recall, derive from simply tracking the occupational whereabouts of 113 workers out of an estimated total of 30,000, and from more in-depth interviewing of twenty-five from that list.

Back on the streets, then, what was the occupational mobility of these 113 workers, over a period of eight to thirteen years? In 2009, seventy-two workers, or 64 per cent of the sample, still worked on *daladala*. Forty-one workers, or 36 per cent of the sample, had moved on.[5] In 2014, the number of workers still working on *daladala* had dropped to fifty out of 105, or 48 per cent of the sample. Such an aggregate figure conceals important nuances in the dynamism of the labour market, whereby the total of fifty workers is an amalgamation of the eight workers who rejoined work on *daladala* after 2009, in addition to the forty-two workers who never ceased to work on buses. In 2014, twenty-two workers of the seventy-two who were recorded to be

---

[5] In 2009 the number of active workers was 113, as three workers had died.

working on *daladala* in 2009 had switched to other types of employment. They followed the earlier move by thirty-three workers, who had switched to other jobs in 2009. Thus in 2014, thirteen years after the initial list was generated, fifty-five workers, or 52 per cent of the sample, had changed employment.

Analysis of these numbers immediately calls into question the mainstream narrative that sees work in the informal economy, notwithstanding its hardship, as leading to steady occupational mobility and career progression, so that informal employment acts like a training centre where entrepreneurs earn the skills on which to draw in later jobs. My research suggests that the opportunity to move on does not easily materialize, and that the occupational immobility of *daladala* workers is significant. In the period from 2001 to 2009, an average of only 5.1 people a year, out of 113, left such employment. For the period from 2001 to 2014, the average further declines to 4.2 people per year. If one were to metaphorically split these 113 workers into four classrooms of twenty-eight, only one classmate a year moves to another job. Nearly half of the workers (48 per cent) from the initial 2001 list were still working on *daladala* thirteen years on. To the extent that workers move out of employment on *daladala*, they do so at a very slow pace. This is in contrast to the aspirations of *daladala* workers for their own progression from employment on urban buses, which, as we have already seen in Dongo's quote, is aligned to the mainstream fantasy of upward mobility.

For the majority of workers, the aspiration for better jobs remains a dream. In the current climate in the Tanzanian economy, the shortage of alternative employment turns work on *daladala*, a 'job to pass the time' by workers' accounts, into a lifelong occupation, and one for which people need to face stiff competition from other job seekers. Once more, *kazi mbaya, ukiwa nayo* ('Bad job, if you have one'). Tragically, then, while *siku zinakwenda* ('Days go by') and *hela ya kula* ('Enough money to eat') (interview with Rama, 2014; interview with Muhidini, 2014) are the two expressions that workers most often use to characterize returns from work on *daladala* and the standards of living they enable, for the majority of this sample thirteen years (in some cases even longer) have become the sum of the 'days that go by'.

So why is work on *daladala* so bad? The reader will be aware by now that earnings for workers in the sector are low and erratic, as they are squeezed by bus owners. But there is more to it. As Dotto, a former worker, explains, the modalities of payment are also significant:

Money from *daladala* doesn't come at the end of the month, so that you can say, with this wage, I will do this. With *daladala* you cannot set goals, today you get money, for three days you don't. (Interview with Dotto, 2014)

Thomas, another former *daladala* worker, further elaborates on how the modality of payment interacts with the nature of work:

> . . . work on *daladala*, its money is like from hell, if you have 50,000 shillings, the day after it has disappeared.[6] The work I do now [driver of a school bus], I get the money all together, differently from *daladala*. If you get money every day, it is not easy to put savings aside. Even if you save money, you might get stopped by traffic police, and you end up eating up your savings. It is a problem, you get money to eat only. The only exception is to get an employer who understands how hard it is, who expects a small sum of money each day [*hesabu*], and who would be prepared to pay for small problems himself.   (Interview with Thomas, 2014)

As I have shown, however, such employers are not easily found. Instead, the vast majority of employers take advantage of the oversupply of labour by squeezing workers with, as we have seen, a resulting negative impact on working conditions and returns from work. Working conditions are unsustainable on two further counts. First, as Rajabu, now a taxi driver, put it, 'there is no respect in *daladala* work. Everyone deprecates you, and everyone thinks of *daladala* as their field to go and harvest from' (interview with Rajabu, 2014). As Rama, another former worker and now a lorry driver, expands: 'That job is hard, here you can relax. In *daladala* [. . .] the owner wants his money, traffic police want their money, the petrol station wants its money, the driver wants his money, and the conductor wants his money. Where is all this money going to come from? There is money but there are too many groups to share it amongst' (interview with Rama, 2014).

In addition to the unsustainable stress levels, the health implications caused by working on buses are an important negative aspect of the job. In the words of Abasi, who became a private chauffeur in 2004 after eleven years of working on *daladala*: 'My body today is very different from those days. You are exhausted on *daladala*. There are days you think, "Shall I get up or not? But there is no one else, so you need to go"' (interview with Abasi, 2014). In sum, work on these buses is characterized by a low and unpredictable income and high stress levels, with negative effects on health.

So what kind of workers stay in this sector and why? What is there in it for them? The analysis now reviews the potted occupational histories of some men who have remained in the sector in search of clues to an answer. Before proceeding, a note is needed about the importance of handling with care the insights of workers reflecting upon different occupational trajectories, in a context of stiff competition for work and daily hardship. Most notably, the comments by those who moved on from *daladala* at times reproduced, with a self-flattering spin, the message underpinning writings such as 'hard work

---

[6] On witchcraft narratives and *daladala*, see Sanders (2008).

pays' and there is 'money if you are smart'. Thus, while all the information gathered was revealing, a large part of the task was to sift through interviews in order to ascertain when particular comments were more an expression of bitterness for not having been one of the lucky ones, or of disdain for those 'stuck' working with *daladala*.

6.2 Histories of Occupational Immobility

6.2.1 *Juma Masuka*

Juma Masuka is the first man whose career I briefly review.[7] The first thing that his trajectory highlights is that the category 'those who remained on *daladala*' needs further unpacking. For while this is an accurate description of Juma's employment history, it conceals his progression from conductor, a job which he started as a teenager in the late 1990s, to driver, not forgetting a considerable length of time spent 'on the bench'. A failed street vendor, Juma came to work on *daladala* through a friend who was a conductor, and whom he worked with on a bus on another route. His connection to the route under analysis (route X hereafter) came from another friend, who was a *daladala* driver and who asked Juma to work as his conductor on a bus called 'Masuka Trans', from which he took his name. Having lost that job, Juma decided to stay at route X, and after some time on the bench, he became the conductor for another driver from route X, Mr Monde, whom he had got to know while at work and 'on the bench'. Mr Monde was the one who taught him to drive. As Juma recalls, 'he would tell me to start the bus, to wait for passengers, to drive up to point Y [a bus stop about 500 metres away from the end of the route], to then park again, until I learnt to drive' (interview with Masuka, 2014). Work as a conductor, with all its hardship, still allowed him to save enough money to obtain his driving licence in 2003, and to support his parents a little, with whom he was still living. With his newly acquired licence, Juma began looking for *day waka* shifts as a driver. First, he worked for a driver of route X, known by the enlightening name of *Zee la kuwawa* (Mister Bankrupt). The relationship with this driver was helpful, on the one hand, but also highly exploitative on the other. 'They were long shifts, from 10 a.m. until the end of work, and then I was being paid less than half of the income, and even being asked to come earlier the day after. He was squeezing me too much', prompting Juma to look and find another driver that would give him *day waka*. After months of this life, relying on a few hours of work on an irregular basis, he obtained his first bus as a driver 'with a livelihood'

---

[7] See section 3.6 for Juma's employment history prior to work on *daladala*.

(*maisha*) in late 2003. This opportunity came through another driver, whom he had worked with as a conductor earlier on in his career. His colleague's employer had bought an additional bus, and was looking for another driver. Juma worked on this bus for about a year until, on 14 October 2004, he was involved in a bad traffic accident. The bus overturned and Juma seriously injured his arm. As the vehicle was repaired before he had recovered, he lost his job.

Upon his return to his *kijiwe* ('pavement'), Juma struggled for over a year to obtain sufficient *day waka* shifts, or to find another bus to work 'with a livelihood'. As the situation was unsustainable, in 2006 he decided to change his 'pavement', which became a stop between Morogoro and Morocco Road. Juma was totally new to that spot, and had no contacts there. It was a stop that he chose for geographical convenience, as it was located at a walking distance from the place Juma rents. As people 'became used to [him]', Juma began to be offered some *day waka* shifts until a 'proper job' became available. A woman from that area, who had a *daladala* and was looking for a driver, gave him his first full-time job at his new stop. She had a contract with a school which hired her vehicle as a school bus, which Juma and his colleague operated for no remuneration. In the evening Juma and his conductor, at their own risk, would then use the vehicle as a pirate *daladala* to earn their daily income as well as the money for petrol. When the contract with the school ended, his employer switched to using the bus as a *daladala*. In 2010, having failed to make enough money to repay the loan with which she had bought the bus, she decided to sell the vehicle. From that day, until the time of fieldwork in 2014, Juma sat on the bench and relied on *day waka* shifts.

This is Juma's trajectory over nearly twenty years of employment in *daladala*: first as a conductor and then as a driver, on and off, incorporating an accident with a major injury, several job losses, a change of 'workplace', and considerable time spent on the bench. It has allowed Juma to save enough money to get married, to rent a place where he now lives with his wife (who cooks and sells cassava, samosa, and doughnuts from home) and his son, who is currently at primary school. To get married and to father a son are important steps in his life, but Juma feels that between him and a better and more secure future stands the problem that 'there is no other employment, more than this on *daladala*. If I can get something other this, I would do it, but as for now I have no job, I do two/three trips, I get some money, and days go by'.

6.2.2 *Uwazi*

Having similarly come to work on *daladala* from a failed attempt at small-scale newspaper selling, Uwazi was a conductor who spent his time as a *daladalaman* on and off the bench.[8] During long periods on the bench, he supplemented his income from *day waka* with other ways of making a living. During my fieldwork spell in 2011, the availability of *day waka* shifts was particularly erratic. In response, taking advantage of his lodging being located in Vingunguti (the area of the city in which slaughterhouses were located), Uwazi would wake up in the early hours of the morning, buy the heads of slaughtered goats, and sell them to the women who cook goat soup. By 2014 Uwazi had come off the bench, and was sharing in equal parts a job with a colleague, so that they worked as *daladalamen* 'with a livelihood' for half a day each. Whilst his income only enabled him to rent a room in a very poor neighbourhood, his wife and two children, who in 2009 were living in the village from where they hail, had now moved to join him in Dar es Salaam.

6.2.3 *Kajembe and Ngaika*

Kajembe and Ngaika were both older men, in their sixties in 2014, and veterans of work in the sector on which they relied for more than twenty years.[9] Their situation underlines the importance of age and ageing in work on *daladala*. Kajembe explained how having worked on several routes meant 'some owners became used to me, they know me' (interview with Kajembe, 2014). But in 2014 he had been relying on *day waka* for some time, and was likely to continue to do so, as employers preferred younger drivers, as they were able to sustain the very tough working conditions better. Ageing seemed to push workers inexorably towards the margin of the *daladala* labour market. Similarly, Rashidi Ngaika (with Kajembe, one of the three workers who acted as key informants in tracking the occupational whereabouts of workers) was on the bench both in 2009 and 2014 (and in between the two visits), and explained how he was increasingly sidelined to the bench by employers who preferred younger drivers to him: 'I can be here for even a week without getting *day waka*' (interview with Ngaika, 2009). At least both these men lived with their families, unlike Sulemani, to whom we now turn.

6.2.4 *Sulemani*

Sulemani was amongst the oldest of the *daladala* workers in town, as he had been working in the sector since the early 1980s, 'when the fare was 5 shillings'. He had been on the bench since 2002, relying on *day waka*:

---

[8] See section 3.6 for Uwazi's employment history prior to work on *daladala*.
[9] See section 3.6 for Kajembe's employment history prior to work on *daladala*.

We sit here, we talk, a life of trouble, deep trouble, you sit with hunger, as you see me today, I haven't got any bus or anything else... These days to work on *daladala* you need someone to take you to the employer, in the old days you went and offered yourself. These days you need someone to hold your hand, 'bwana this is my relative, I know him', and the boss needs to know him, so that if you disappear, if you do a mistake, it will be easy to get you.   (Interview with Sulemani, 2009)

Sulemani seemed stuck in Dar es Salaam, while his wife and two children were in Tanga, from where he originated:

I can go and visit when things are going well. How can I go when I haven't got even the money for breakfast? I will have to go and see them with enough money, not with 10,000 shillings. So you need a job, 100,000 shillings at least, to go there. The money for the bus ticket to and from Tanga, clothes for my parents and family, and enough money to use while I am there. How can you get this money with *day waka*... We live like birds. Actually a bird is better off as he knows that he will eat. There is no way out.   (Interview with Sulemani, 2009)

Sulemani, in a common escapist move, laid to sleep what seemed like unbearable thoughts about existential failure by heavy drinking, which further deepened his problems.[10]

6.3 Histories of Occupational Mobility

What about those who moved on to other jobs? Of the fifty-five workers who did so, the vast majority, forty-two, worked in transport. More specifically: thirteen were hired by companies as drivers of vans or office cars, eleven were lorry drivers, eight were chauffeurs, eight were taxi drivers, and two worked as labour overseers of individuals who owned fleets of *daladala*. The above were individuals who built on the skills acquired and/or honed through *daladala* work and who subsequently applied them to other driving jobs. The remaining thirteen workers undertook a range of jobs/activities, such as farmer (three), change seller (two), mechanic, casual worker in a car wash, security guard, unloader in a supermarket, shoe shiner, witch doctor, waiter in a small restaurant, and carpenter. The potted occupational biographies of some of these workers help to capture the motivations and circumstances that allowed some to move on from employment on *daladala* to other jobs and their reflections on how their new occupation compares to the previous one.

---

[10] On alcohol drinking among the poor in India, see Breman (2010: 137). On alcohol and drug consumption among *daladala* workers, see Kisyombe (2005) and UWAMADAR (2010).

6.3.1 *Rajabu*

Rajabu was 38 in 2014.[11]A chauffeur who lost his job in 1998, he found employment as a *daladala* driver through his uncle, who connected him to a bus owner whose bus was operating on route X. After four or five months, that job came to an end, as his employer was unable to pay for the maintenance and repairs of the bus, which he sold. By then Rajabu had developed some track record of work on *daladala*. 'I had already done this job, and people knew that' (interview with Rajabu, 2014). One of his neighbours became his next employer: 'I have a bus, interested?' In taking on the job, Rajabu also convinced the owner to shift the route on which his bus operated to route X. After five or six months, a better opportunity presented itself, in the shape of an employer who not only intended to buy a brand new and therefore more reliable bus, but also offered Rajabu a wage at the end of each month, on top of the daily return that could be earned after delivering the employer's daily sum. Luckily for Rajabu, the daily sum that this employer demanded was at the market rate and not significantly above it, as was more common in situations in which a monthly wage was paid. It must be noted that such employers are not the norm in the *daladala* sector. When the bus registration was completed, Rajabu started work, having 'handed over' the job on his old bus to a driver who was an old friend of his and on the bench at that time. 'I knew he would do a good job, I trusted him.' Rajabu worked with this new bus until the bus 'became tired' and was sold. Rajabu was back on the bench, relying on *day waka* for two months, until the next twist in his career when he became a taxi driver in 2003. His shift to taxi driving came about because of his good relations with his previous employer. When the latter reinvested part of the profits from the sale of a *daladala* in a taxi, he chose Rajabu as the driver. Having worked on that taxi for a year, until it was sold, he found a new employer and many others after that. Reflecting on how working on a *daladala* compares to working in a taxi, Rajabu's first point was about the lack of respect that *daladala* workers command: 'Everyone deprecates you, and everyone thinks of *daladala* as their field to go and harvest from. So I prefer working in a taxi.' But it was also a type of work that suited him at a time in which he had no wife or dependants, and from which he was able to move on to better things.

> I could earn money to eat and a bit of savings. With taxis, my economic conditions have improved a bit, I achieved things, I got married, and I built my own place. But I cannot say I will not go back to *daladala*. The prisoner doesn't choose his prison. I have a network in *daladala*, I have my friends, *day waka* in Y, in Z, they know me. It is a challenge, it is a step of life, I had enough of that experience. It taught me to be sharp, the head hurts, I did it, I have seen it.   (Interview with Rajabu, 2014)

---

[11] See section 3.6 for Rajabu's employment history prior to work on *daladala*.

6.3.2 *Abasi*

Others are not prepared to work on *daladala* again, pointing, above all, to the unsustainability of its working conditions. Take Abasi, 38 years old in 2014, who started off as a conductor at the age of 17, in 1993, and then became a driver in 2001. Since 2004, he has worked as a chauffeur. As he put it: 'I won't be able to work again on *daladala* as a driver, it is very hard, and it tires you badly. I started work at 4, back home, if early, at 9 or 10, to sleep at 12. It is not good for your health' (interview with Abasi, 2014).

By comparison, his new job has the advantages of allowing him time to rest at the weekend, and a formal contract with social security attached. What is interesting about Abasi's occupational trajectory is not only what it reveals about its direction, but also the sobering reminder it gives about how careful one needs to be when handling informants' insights as 'evidence'. Abasi's story throws new light on Kajembe's one, presented earlier in this chapter, through details that did not emerge in discussions with Kajembe himself. Abasi was working as conductor with Kajembe, on route X.

> Kajembe had the habit of lying to the owner, to tell him that his bus had broken down. But once he was found out. Kajembe took the bus to a garage to fix the door, but unfortunately for him it was a garage where a relative of the owner worked. So when the owner asked him how much did it cost to fix the door, and Kajembe told the figure of 15,000 shillings, when it actually cost 6,000 shillings, he lost his job.

Kajembe's fall was Abasi's fortune as it opened the door of driving for him. At that time, Abasi had a class B licence. While this did not legally allow him to drive *daladala*, the owner of the bus turned a blind eye. Such a choice, Abasi suggests, came from the owner being impressed by Abasi's hard work ethic as, whenever he checked on his *daladala*, he found Abasi working on it as a conductor (as opposed to taking time off by letting a colleague from the bench take over for part of the day). Abasi worked as a *daladala* driver for five years, a profession which he found unsustainable but, unlike most of his colleagues, rewarding. 'There is money in *daladala*. It is up to you how you use it. If you go and use it [with the presumption that] "tomorrow I'll get it again", but that might not be the case.' Abasi mentions the fundamental role that his partner, and now wife, played in looking forward and in managing earnings from his work. 'If you have a woman who is very wise at building life "20,000 shillings, let's put them aside, 5,000 shillings for food". My wife helped us a lot because when I came back from work I just handed her my daily earning. I had breakfast at work, lunch at work, only for supper I ate at home.' It is from work on *daladala* that Abasi found the resources to get married and buy a plot of land on which he is slowly building a house.

Such work was, however, physically unsustainable and high risk. The opportunity to move on from it first came from his wife's connections. She was a teacher at a school, and Abasi became a driver of one of the school buses. The new job was an improvement above all because of its working hours and conditions, 'with time to rest'. As such, it was sustainable, and furthermore it came with access to social security 'which will help me later, if I am ill, and my children, when I die. In *daladala*, you are on your own'. The opportunity to move to an even better job came from a very good old friend from his days on *daladala*, Frenky Nyanda (whose name also appears on the workers' list), who in 2014 worked as a taxi driver. Having heard that the wife of the Education Minister, Kawamba, was looking for a driver, Frenky forwarded this information to Abasi, who put himself forward and subsequently began work for her, a job which he still undertook in 2016.

6.3.3 Dotto

> I don't work on *daladala* any longer. These days it is too hard, it doesn't make any sense. It doesn't pay. Work on *daladala* depends on how the owner is. With a big *hesabu* you can't. With a smaller *hesabu* it is better. For workers to share what is left, may be 20,000 shillings, you woke the car up at 4, until 9 p.m., what is this about? It is therefore better to look for employment with someone, maybe 200,000 shilling per month, but from 6 a.m. to 7 p.m. only (sic). And when I finish work I get time to rest, I get holidays, unpaid, but I rest. Resting the body, with *daladala* there is no day off. Good jobs in *daladala* is for very few, good employers are rare.   (Interview with Dotto, 2014)

Dotto has a long history with *daladala*. He started as a *mpiga debe*, the job of those who fill buses with passengers, back in 1993. Over time, Dotto established himself, together with Asenga (who will feature in section 6.3.4) as *mkuu wa reli* ('the head of the ' "railway line" '), as one of the leaders of the route. I have already discussed the dangers and low status associated with work as *wapiga debe*.[12] Dotto himself was the informant who stated:

> Any man with a sound brain knows that shouting a destination and pulling people into the bus all day is not a job. We do it because we are in trouble. But it is not a job ... The heart hurts when you think about life, because it is not life to be here at the station. You can't bring your family [and say] 'Come to the office' ... This is a pavement and as it is a pavement it is not an office.   (Interview with Dotto, 2014)

As one of the heads of the 'railway line', Dotto played a major role in determining who was entitled to its income-generating work shifts, as well as in ensuring that those who were entitled actually had access to them. Given the severe need for cash among its members, being the head of the 'railway line' required, in the words of Dongo, a former *daladala* worker colleague, 'being

---

[12] See section 4.4.3, for a discussion of *wapiga debe*.

tough, fair, being able to talk and listen, and being able to handle problems when they come' (interview with Dongo, 2010). Through this work, occasional shifts as *day waka*, and becoming one of the leaders of 'the railway line', he earned enough money to get a driving licence and become a driver. Tired of *daladala* work, for the reasons outlined above, through a friend he signed up with a labour agency that took him as a driver for a year, to then supply his services to offices that required drivers. His new job paid 250,000 shillings a month, fed him breakfast and lunch, and had better working hours, from 6 a.m. to 5 p.m. But Dotto still faced a very uncertain future as his one-year contract was about to expire at the time of my fieldwork in 2014.

Reflecting on his career, and where it has taken him over twenty years, Dotto's words should be heard by all those writers who unduly celebrate the inventiveness and agency of the urban poor:

> Cleverness without results is pointless. If I was hired by a company, I would turn things around. If I were paid 500,000 shillings, I could save 100,000 shillings each month. In ten months, I would have 1 million, with one million a *bodaboda*, and with earning from it you would find that in one year you get a plot.   (Interview with Dotto, 2014)

Instead, Dotto lived with his wife and three children in rented accommodation. He bitterly complained that 'you can't live life without savings. A funeral, my father is ill, you can't live life without savings. I am 15,000 short, I call Asenga. He is not there, or he is there but he has no money.' One thing Dotto was grateful for, however, was that he could support his children through education, past primary level. 'I believe that even if my life has been bad, for the fact that I made them study it can be better for them.'

6.3.4 *Asenga*

In 2014, Asenga was a 44-year-old man from Rombo in North-Western Tanzania, married with two children. A failed small-scale trader in second-hand clothes, Asenga became a *daladala* conductor in 1990, invited by a friend of his who was a driver.

However, a bad crash while at work prompted Asenga to return to trade in second-hand clothes, which he did from 1993 to 1997. As, again, it did not go well, he went back to being a conductor on *daladala*, this time with a friend on route X. In 2002, health problems forced him to step down from work on *daladala*. Once he recovered, he decided not to return to work as a conductor, and instead he focused on *kupiga debe*—that is, the job of filling buses with passengers. Over time, Asenga established himself, with Dotto, as one of the leaders of route X 'railway line'. Through this, within about one-and-a-half years, Asenga was able to save enough money to start a small kiosk at the station (see Figure 6.1). There he sold newspapers and a few stationery items.

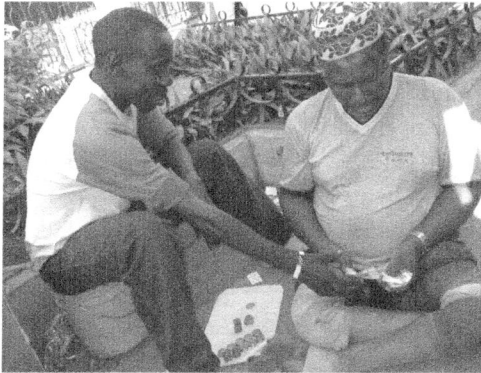

Figure 6.1 Asenga and his kiosk

As business at the kiosk proved sustainable, he called his cousin from Rombo to man the kiosk, while he continued to be the boss of the station and one of its main *wapiga debe*. Through this double source of income, over time Asenga was able to expand the range of goods in which he traded, including juice, water, and biscuits, and increase the size of the kiosk, adding one more table. Such trade in drinks and biscuits proved short-lived, as the increase in the size of his stock of goods made him more vulnerable to the threats of confiscation or bribery attempts by City Council attendants. Instead, as mobile phones and mobile banking spread in Tanzania, Asenga became an Mpesa agent, a seller of mobile phone vouchers as well as providing a station where people could recharge their phones. Furthermore, every day, as rush hour neared, Asenga would take out two very large plastic bags from which he lined up several piles, each worth 1,000 shillings, of 100 or 200 shillings coins. This was because Asenga also sold change to *daladalamen*, who could not afford to slow down their operations in the search for change, and therefore bought 900 shillings worth of coins in exchange for a 1,000 shilling note. From where did Asenga get all these coins? The beggars were his suppliers. One of them, Adamu, lost his legs in a horrific accident in an upcountry bus in which many people lost their lives. Invalided since then, he made a living by begging at *daladala* station Y. As Adamu explained, 'I have been bringing my coins to Asenga for 12 years' (interview with Adamu, 2014). He gave the coins he earned every day to Asenga, who did not make a profit on this exchange, but instead gave Adamu the same value of money back in notes—13,000 shillings on the day I Interviewed him. Why did Adamu bring his coins to Asenga? 'Because the day I don't get anything he looks after us'. It is through this relentless trade with very small margins of profit, six days out seven, sixteen hours a day, that Asenga was able to sustain his family, and to buy a plot of land, on which he had begun to build slowly, and where, if 'the area develops nicely', he planned to move one day.

6.3.5 *Mudi and Kulwa*

It would be misleading and simplistic to read any transition away from *daladala* to other jobs as a success story. Although this did seem to be the aspiration of workers upon entering the sector, and one that was fulfilled by some, as I have shown, another modality of exit could be observed. It had more to do with distress and with being squeezed out of the system. As work in the *daladala* industry had a tendency to increase the number of workers temporarily out of work and 'on the bench' exponentially, the competition between them for fragments of work from overemployed workers 'with a livelihood' was severe. There were only so many people whom the bench could accommodate. Some ultimately faced marginalization from the dearth of day work that was available. Take Mudi Abasi, the brother of Abasi, whom we encountered earlier (in section 6.3.2). Mudi joined his brother, working as his conductor on route X in 2004, as his small-scale business as a carpenter did not pay off. The brothers' work partnership was not long-lasting: shortly after it began, Abasi moved on and became a chauffeur, leaving Mudi behind. Mudi kept one eye open for work as a carpenter, and occasionally his carpentry teacher hired him as a daily labourer. The other eye was trained on getting some *day waka* from the bench. But he was not doing well on either front. 'I look at earnings, carpenter or *daladala*, which one is better. As I am in trouble, as I am here in town to look for money, I try every possibility. Those in trouble have to struggle. In *daladala* I have no bus 'with a livelihood'. On the side of carpentry there is no work. I am just surviving, I have no certainty' (interview with Mudi Abasi, 2009). Mudi was no longer on the bench during fieldwork spells from 2011 onwards.

For some workers the fact that workers like Mudi were brought into *daladala* work by family members, and were then left struggling to find sufficient work, was no coincidence. As Tolu put it, 'a person who has been brought here by someone relies on that person. "Bwana I know Matteo, take him, give him work" ... The person that came here on his own, with the bus of his employer, he came on his own.' True as it might be that workers brought to work on the route by relatives depended on them, it is important to note that Tolu was one of those who 'came on his own', and there is therefore the possibility that his was, above all, a derogatory remark against those he saw as getting a foot into *daladala* work. Furthermore, those who were brought to work on buses by relatives or friends were not the only ones who struggled. Take Kulwa, an old timer of route X, who came to the route as a driver 'on his own' and who, having lost employment on his bus 'for a livelihood', subsequently relied on *day waka*. Similar to Tolu, in 2009 Kulwa boasted proudly about how well-established he was on the route: 'I am an old timer here. I will not miss out on *day waka*. If I move I will get less *day waka*' (interview with Kulwa, 2009). However, by 2014 Kulwa had moved to another route (Mwenge/Tegeta), where he could find more *day waka* than in Masaki.

6.4 Workers' Trajectories: Predictable?

What have we learnt about occupational mobility and immobility by tracking the trajectories of 113 *daladala* workers over thirteen years and by interviewing some of these workers on their work histories? This study was animated by an interest in understanding whether and how individuals exert agency in negotiating the same occupation, and the same structural position in the labour market. Through pinning down and empirically documenting workers' occupational mobility/immobility I have also engaged with 'all structure' and 'all agency' narratives on the African city and on economic informality. My findings suggest the urgent need to move beyond the undue optimism which infuses so much writing on African cities. Above all, evidence of the occupational trajectories of *daladalamen* exposes the fantasy of conceptualizing work in the informal economy, notwithstanding its hardship, as an entry point into other, and better, jobs. In 2009, eight years after the first observation, 64 per cent of workers still worked on *daladala*. In 2014, thirteen years on, just under half of workers, at 48 per cent, were still working on *daladala*. Over the thirteen-year period, an average of 4 per cent of these workers left employment on *daladala* every year. Thus, for far too many workers the ambition to work on *daladala* as a 'passing-the-time' job remained a dream.

Fluidity, unpredictability, and any other word that suggests uncertainty are helpful to describe the circumstances faced by these workers, but it would be highly misleading to read agency of the working poor into these trajectories without attention to structural forces. There are some predictable problems that workers face because of them, such as not knowing whether tomorrow they will be in the bus all day or on a bench. Such uncertainty comes about, first and foremost, because of a structural problem of lack of jobs in the Tanzanian economy, and because of employers' strategy of dumping most of the business risks on the workforce, a strategy which they can promote due to the oversupply of job seekers vis-à-vis demand. *Kazi mbaya, ukiwa nayo* ('Bad job, if you have one').

Attention has been paid to disaggregating across and within the four occupational trajectories. Among those workers that did not move to other jobs by 2014, the single most important marker of difference within this group was age. Older workers experienced a lack of competitive edge compared to younger and more energetic workers whom employers preferred. Squeezed like lemons after years of gruelling work on *daladala*, with family burdens that tend to be more significant, such workers were unable to match the energy levels and endurance of younger colleagues and increasingly spent time on the bench. I have also noted that the vast majority of workers made a transition from conductors to drivers.[13]

Understanding the significance of the trajectories of those who did move on to other jobs must also start from disaggregating the group, as this, together with workers' own perceptions, reveals that it would be simplistic to read 'exit' as 'success'. In some cases, those who left did so because of having become

---

[13] Attempting the transition from conductor to driver came with the risk of being caught in between the two professions. One could fail to establish oneself as a driver whilst no longer being 'seen' as a conductor by colleagues, with the resulting lack of opportunities to work in either position (interview with Kudo Boy, 2014).

marginalized, as the intense competition over fragments of work confined some, inexorably, to far too much time on the bench, and ultimately forced them to move on. Some workers within this group genuinely did move to better jobs, in terms of higher remuneration, better working conditions, and lower stress levels. They tended to transfer driving and transport skills to other subsectors, such as taxi (Rajabu) or lorry driving. Noting the slow speed at which such transitions took place, and that stories of this kind were frequent but not the norm, the only other insight one can gather is that personal connections—and circumstances that are not easily replicable—proved crucial in allowing some workers to move to other jobs. One worker found work as a driver at the school where his wife was a teacher (Abasi). Dotto and Asenga emerged as leaders of the 'railway line', out of over 100 other workers of the same route. This position, which was by its very nature unavailable to many, was used by Asenga as a springboard to small-scale trading and by Dotto to help him to become a driver. Another worker's escape from *daladala* came from his working for a good employer, who gave him a wage and a brand new vehicle on which to work (Rajabu). When this employer then went on to invest his money from *daladala* in the taxi industry, he chose the worker as his driver. Attention has also been paid to examples of workers' improvisation and survivalism, which led some to literally invent jobs where there were none (such as the job of filling buses), but as one of these very workers warned us, 'cleverness without results is pointless'. Beyond these differentiating factors, it would be questionable to stress the significance of other alleged predictors of success and failure to move on to other jobs, such as work ethic, being streetwise, or years of experience. As has been shown, they proved to be more of an insight into the internal tensions that divide the workforce rather than useful entry points to understand its occupational trajectories.

For good or for bad, at the time this book went to press, employment on *daladala* was scheduled to shrink dramatically as the buses were to be relegated to the margins of the city of Dar es Salaam. A new idea, backed up by influential and well-resourced institutions, had been heavily promoted to build 'a better city for better times'. Its genesis, its driving forces, and the twists and turns that it brought to the political economy of public transport in Dar es Salaam are the subject of Chapter 7.

7

The New Face of Neoliberalism

The Bus Rapid Transit Project in Tanzania (2002–16)

7.1 The Political Economy of BRTs

Previous chapters have shown that the principle of 'rolling back' the state has informed nearly three decades of policymaking over public transport in Dar es Salaam, from 1983 to 2014. Close examination of the sector traced the transition from public to private service provision and the progressive deregulation of the activities of bus private operators. At the same time, since the very late 1990s, when the process of liberalization and deregulation reached its peak, Tanzanian public authorities attempted to reclaim some policy space through a number of initiatives on passenger transport matters, ranging from ad hoc interventions to the establishment of a new transport regulator. These regulatory efforts, notwithstanding differences between them, brought two sets of issues to the fore. First, such initiatives revealed the politics of 'actually existing neoliberalism', its tendency to 'self-combustion', and the Tanzanian authorities' relative willingness to intervene to address the tensions that the rolling back of the state had generated. These led to the second set of issues, as such interventions exposed the lack of capacity of the state to both formulate and, above all, to enforce regulation of the activities of private operators. The gap between the magnitude of the city's transport problems and the state's capacity to deal with them appeared macroscopic. Donor-imposed fiscal austerity and the shrinking of the size of the public sector—the defining features of this first phase of neoliberalism—were visible in the dearth of the human and financial resources available to public transport institutions in Dar es Salaam.

This chapter, however, charts the subsequent changing face of neoliberalism in public transport in Dar es Salaam, signalled by a remarkable departure from the previously described period of limited government capacity and fiscal austerity. In 2016, a large-scale initiative to radically change the transport system was launched. This was the Dar es Salaam Rapid Transit project (DART), the Tanzanian version of Bus Rapid Transit systems (BRT). To facilitate it, a substantial and well-resourced unit was established within the Prime Minister's Office with funding from the World Bank. The capacity of the public sector on transport in the city has thus been significantly boosted, but to what end? Stepping out of Dar es Salaam for a moment, such a shift conforms to the broader transition from a first phase of 'roll-back' neoliberalism, to a second in which state intervention is more actively deployed both to manage the tensions generated by the first phase of neoliberalism and further promote the interests of private capital. 'To put it crudely, once you have done as much privatization as the system will bear under the neo-liberal rhetoric of withdrawal of state intervention, then the time has come to use the state to correct market imperfections and to improve its workings, as in Public–Private Partnerships (PPP)' (Van Waeyenberge et al. 2011: 9). As BRTs are promoted through PPPs in urban transport, the above characterization of the changing face of neoliberalism pertains to the direction of policymaking in Dar es Salaam.

Bus Rapid Transit systems have been increasingly promoted as the solution to chronic and rapidly escalating traffic congestion and to the low quality of public transport provision, widely shared traits of urban life in developing countries today. In 2007, forty cities across six continents had BRT systems (Wright and Hook 2007: 1). By March 2016, the figure had risen to 202.[1] Operating BRTs in Africa include those in Johannesburg, Cape Town, and Lagos. In Dar es Salaam, the BRT is currently at an advanced stage of construction. Accra, Addis Ababa, Dakar, Kampala, Nairobi, Maputo, and Luanda are amongst the African metropolises where preparation for BRT is at an early stage.

'Think rail, see bus' goes the BRT motto. Proponents of BRT systems stress how they combine the flexibility of bus transit with the benefits of a rail-based mass transport system (namely speed, reliability, and mass ridership), at a fraction of rail's costs.[2] However, while BRTs are to some extent context-specific, there are five common characteristics that help to explain what is at stake in their promotion. First, while cheaper than rail systems, BRT systems still require substantial investment. International finance led by the World Bank has played a pivotal role in providing the funds for their implementation. Second, although BRTs are publicly funded, a conditionality attached to World Bank lending is that private companies operate the buses. The public sector's role is to oversee the system and carry out quality controls on the

---

[1] On the current figure see 'Global BRT Data' <http://brtdata.org/#/location>.

[2] ITDP estimates that BRTs are four to twenty times cheaper than tram or Light Rail Transit (LRT) and ten to 100 times cheaper than metro (Wright and Hook 2007: 1).

service providers. Therefore, although PPPs are not necessarily the only possible institutional set-up for BRTs, PPPs are de facto the way in which most BRTs operate, as suggested by their promoters. For example, the ITDP's planning guide for BRTs states that implementing BRTs through PPPs is one of the core principles of the 'effective BRT model' (Wright and Hook 2007). BRTs are therefore to be understood as the urban transport expression of public–private partnerships, the rationale for, and benefits of which, are contested (Loxley 2013). Third, a major characteristic of BRTs is that they entail the phasing out of privately owned minibuses from the main arteries of public transport systems, and their deployment on feeder routes. As BRT buses are new and less polluting than those which provide public transport in many developing countries, advocates of BRT outline the environmental and traffic reduction benefits of bus-switching. Fourth, BRT delivers faster trips thanks to off-board fare collection, platform-level boarding, and a fundamental shift in the rights to urban road use, as BRT buses are normally granted two dedicated lanes. Finally, as a result of this, BRT systems require major upgrading of the urban road infrastructure, including the rebuilding and widening of main roads.

Research suggesting that BRT successes in Latin America opened 'a new era in low-cost, high-quality' transport is key to the argument for BRTs in Africa (IEA 2002). Described as the 'world reference point for bus rapid transit systems' (Quality Public Transport 2012: 1), Bogotá's highly celebrated Transmilenio became the 'first mass transit system in the world designated as a Clean Development Mechanism under the UN Framework Convention on Climate Change' (ITF 2010: 1). At the 2012 UN Sustainable Development Conference in Rio de Janeiro, international development banks pledged US$175 billion over ten years to support sustainable transport in developing countries (WRI 2012), with BRTs playing 'a key role in creating sustainable [urban] futures' (Cervero and Dai 2014: 128). Furthermore, international development funding has been channelled into building links between Bogotá and prospective BRT systems, resulting in officials from more than twenty countries visiting Transmilenio to learn from its (alleged) success.[3] As transport systems that benefit the economy as well as the environment and the poor, BRTs are championed as the ultimate 'win–win' intervention to solve public transport problems.

In contrast to such positive accounts, this chapter takes a critical look at BRTs through a case study of DART. The object of analysis is the remarkable delay in the implementation of DART and how these delays can be understood as a consequence of the politics and resistance out of which 'actually

---

[3] On learning about BRT from Bogotá, and more broadly from South American cities, with the aim of implementing it in South African cities, see Wood (2014, 2015).

existing neoliberalism' emerges. Selected as the BRT forerunner in Africa in the early 2000s, the imminence of DART has been announced time and again over the years. In 2004, and again in 2007, the project's implementation was promised to no avail (*Mtanzania*, 12 March 2004, 8; *The Guardian*, 23 January 2007). In June 2008, DART's Chief Executive declared that work would start in September and that 'all things being equal, the project should become operational by 2010' (*The Guardian*, 2 June 2008, 1–2). In 2011, as the Dar es Salaam Commuter Bus Association (DARCOBOA) Chairman remarked, 'they [were] even embarrassed to mention a time for completion' (interview with Mabrouk, 2011). Construction work finally gathered momentum in 2013. In May 2014, an estimated 60 per cent of the construction was completed, with completion scheduled for the end of 2015 (Rebel Group 2014: 10). Three months into 2016, however, the project had not yet started to operate.

A study of the delays with DART is important as it reveals the contradictions intrinsic to BRTs in one particular context. This requires an understanding of the political economy of DART: what different Tanzanian actors stood to lose from its implementation, and the way in which they were able to resist and influence the project. Indeed, as Gordon Pirie puts it, 'the politics of BRT are delicate: strong leadership is needed to deal with the vested interests of informal paratransit operators' (Pirie 2014: 136). Such politics, which I will argue underlie the lack of leadership behind DART, are at the centre of this chapter. Alongside domestic politics, the chapter also analyses a set of vested interests behind BRTs, and in particular one that is far too often lacking scholarly attention: that of international finance and its NGO brokers who adamantly advocate for their adoption.[4]

The chapter starts by analysing the international lobby that promotes BRTs around the world, and its role not only in financing BRTs but also in generating evidence about global BRT experiences. Attention is paid to the manufacturing of Bogotá's BRT experience as a success story, the shortcomings of which are exposed. Following this, the chapter considers Dar es Salaam's experience of BRT. Drawing on policy documents, newspapers, and interviews carried out during fieldwork in 2010, 2011, and 2014, the analysis highlights the proximate and deeper causes of delays with the project. It is important to note that the type of evidence that it was possible to collect for this chapter was shaped by the political nature of DART and the ambiguous stance of Tanzanian leaders towards it, as most respondents were keen to avoid open confrontations. Public authorities, for example, displayed a strong commitment to DART in words and yet a failure to act accordingly. Making sense of

---

[4] It is interesting to note that Hidalgo and Gutiérrez (2013: 9) exclude lobbying by international finance as among the 'reasons behind the explosive growth of BRT'.

this ambiguity required piecing together fragments of at times conflicting evidence, which on their own were hardly conclusive. The process started with press coverage of DART, as this gave clues as to the main issues that stalled its progress. These clues were further explored through the study of policy reports and interviews with key informants. The interviews were central to grasping the existence of incompatible versions of the same events, the tensions associated with the project, and how they played themselves out in the bigger picture of contemporary Tanzanian politics. This required, as with all research, that I arm myself with scepticism when considering the reliability of the sources and the possible agendas of the informants.

Based on a careful and contextualized interpretation of these sources, the chapter explores the hypothesis that the delay in implementation, far from being the consequence of simple project management failures, is rooted in the politics of BRTs. Precisely because DART cannot be meaningfully understood as a 'win–win' intervention, and there were constituencies in the transport system that stood to lose from its implementation, DART posed dilemmas to the Tanzanian government. On the one side stood the World Bank, funder of the project. On the other were a number of local actors, such as the current paratransit operators, their workforce, the Dar es Salaam City Council (DCC), and those who own property that stood to be expropriated, and compensated for, by DART. Each of these actors was compelled to make way for the project and, as a result, attempted to resist it. In addition, there were prospective operators who tried to force their way into the project, seemingly against the will of its planners. Thus, the story that follows traces the precarious balancing act performed by the Tanzanian political leadership, which ended up caught between a rock and a hard place. It analyses the ways in which the government negotiated the agendas of various local constituencies, to some of whom it was electorally accountable, alongside that of the World Bank.

The relationship between the government, the ruling party, and voters is particularly important to this story. Research on Tanzania's political landscape highlights the way in which elections have motivated the ruling party, *Chama cha Mapinduzi* (CCM) (Party of the Revolution), to adopt policies likely to win large numbers of votes by virtue of their popularity with voters, their close association with the ruling party, and their countrywide reach—such as the reintroduction of subsidies for fertilizers (Kjær and Therkildsen 2013: 592–614). The electoral competition faced by CCM also affected the modalities of competition within CCM and the overall functioning of the party, with an unhealthy tendency towards short-termism and corruption. Kelsall suggests that 'party roots and branches must be generously fed and watered' from the top to prevent junior cadres from defecting to opposition parties, and to influence their choice of party presidential candidate (Kelsall 2013: 61).[5] The increased prominence within CCM of business magnates like Iddi Simba and Rostam Aziz, who acted as campaign managers for two former Tanzanian Presidents (Jakaya Kikwete, from 2005 to 2015, and Benjamin Mkapa, from 1995 to 2005, respectively), was also linked to political competition and to the increasingly important role played by the resources of wealthy men in securing the 'support of local activists in primary elections'. The stance of the CCM leadership towards corruption seemed ambivalent. On the one hand, important successes have been registered in fighting corruption, most notably through expenditure reforms in public finance. On the other hand, continuous corruption scandals involving senior figures, such as the 2005–8 Tanzanian Prime Minister Edward Lowassa, indicate that the CCM centre lacked the ability to rein in 'networks within the party who supported deals that were manifestly bad for the interests of the country and indeed for the leadership' (Gray and Khan 2010: 352).[6]

Such general reflections on the electoral implications of policies adopted by the CCM and the roles played in the project by Edward Lowassa and Iddi Simba—controversial characters in Tanzania's recent political history—are useful in understanding the vicissitudes in the implementation of DART. More specifically, these considerations point to the mixed motivations of the CCM towards the BRT project. As I will show, while the government hoped to benefit from the additional popularity that a successful initiative might bring, DART also presented a huge legitimacy challenge to the ruling party, adding another chapter to the growing, if long-standing, debate on the extent to which Tanzanian interests have been guarded by its government when dealing with foreign investors in the post-socialist order of things (Lange 2011; Ponte 2004; Aminzade 2013).[7] Key players were also willing to manipulate the programme for private benefit, while others feared that the CCM might lose support if DART's implementation had negative effects on the hundreds of thousands of daily passengers using public transport and on pre-existing operators. It is this complex set of motivations that explains the tepid commitment to BRT at the heart of government, and hence its slow pace of implementation.

---

[5] In 1992 the party's procedure for selecting the party presidential candidate was democratized so that since then delegates vote for their chosen candidate at the party's national conference.

[6] See Gray (2012) on the resignation of Edward Lowassa in 2006. This was caused by the infamous electricity supply scam by Richmond-Dowans Limited, a company that successfully bid for the supply of electricity to TANESCO, the state-owned electricity company, despite having no previous experience in the sector and which, unsurprisingly, failed to deliver. When it became public, in 2006, that its contract guaranteed Richmond-Dowans Limited the hefty fee of US $100,000 per day without even the supply of electricity and that the company was fictitious, Lowassa was forced to resign due to his role in brokering the deal.

[7] See section 1.5.

7.2 The BRT Evangelical Society

Looking closely at research on BRT and the narrative that portrays it as the solution to urban transport problems in developing countries, one realizes that a tightly knit web of institutions lies behind. Who the main actors are, and what their stakes might be in BRTs, is therefore an important part of the story. The World Bank is the key player, for it provides not only the loans to make BRTs happen, but also funding for most of the research on them. Another important actor is Volvo, which—not coincidentally perhaps— supplies buses to many BRT systems. It also supports 'Across Latitudes and Cultures—Bus Rapid Transit', which is the BRT Center of Excellence, whose members include four academic institutions and EMBARQ.[8] EMBARQ, which is the World Resources Institute's Initiative for Sustainable Urban Mobility, credits itself for having 'played a major role in expanding the BRT concept to cities throughout the world' (WRI 2012) and is one of the organizations behind 'Global BRT Data', the most up-to-date dataset on BRTs.[9]

Another of the main actors, the Institute for Transportation and Development Policy (ITDP), a Washington-based NGO, is unfailingly on the horizon wherever BRTs are implemented. ITDP's growth from a small advocacy NGO to an organization with over sixty staff members in offices in Africa, Asia, and Latin America has been associated with access to BRT funding. ITDP has played different roles in this capacity. It produced a BRT planning guide, carried out prefeasibility studies in various cities, signposted potential new sources of funding for BRTs, and has been at the forefront of studies on BRT impacts (Matsumoto 2006).

In 2011, the ITDP Board of Directors, worthy of scrutiny as it provides some indication as to whom the NGO is accountable, included the managing director of the Goldman Sachs Urban Investment Group, a representative from the world giant investment firm Carlyle Group, and two representatives of the World Bank, including a retired former Vice President of the Bank (ITDP 2011: 24). International finance obviously has huge stakes in the opening up of urban public transport markets—and more broadly of public utilities markets—in developing countries, and in the funding of the infrastructural work that they require.[10] These are examples of the vested interests of the institutions that present Bogotá's Transmilenio—and BRT more broadly—as a success.

---

[8] See 'Old BRT "Members"', n.d. <http://www.brt.cl/about-us/members/>, accessed 31 January 2017.
[9] See n. 1.
[10] See Hall (2014: 5–44) for an excellent overview of corporate interests and networks in promoting PPPs.

By contrast, independent research and media coverage on Transmilenio present a more ambivalent picture, in which the positive impacts of BRT coexist alongside its negative consequences. Travel times and the quality of transport improved with Transmilenio, but claims that it is 'providing reliable transport accessibility for the poor' (World Bank 2009) sit at odds with increases in transport fares, a trend observed elsewhere (Hidalgo et al. 2007; Munoz et al. 2013).[11] Increased fares have inevitably prevented the poor from accessing the service and have led to public protests demanding lower fares (BBC News 2012). Such demands could not be met, as the bargaining power of the public regulatory body vis-à-vis the private tenders was low, as is often the case with public–private partnerships. Furthermore, the inclusion of previous public transport operators often proved problematic. In Bogotá, ownership of BRT buses increasingly became concentrated in the hands of a few private operators (ITF 2010), while other contexts presented their own distinctive tensions with inclusion, at times violent (Paget-Seekins et al. 2015; Walters and Cloete 2008). Another major problem with Transmilenio is the contraction in employment opportunities that accompanied the higher productivity of labour. The proclaimed goal of replacing the exploitative informal employment relations of the pre-existing transport system with better, formal jobs ran into difficulties as workers faced new types of pressures from employers under the new system. Access to these jobs for those who worked on the previous system was not straightforward. 'Only one in seven of the bus drivers in the old system were able to find work in the new one', partly because the harsh working conditions of the previous system prevented many from passing the medical test for a job in the new system (Quality Public Transport 2012; Porter 2010). Although the World Bank has praised Transmilenio as a 'financially self-sustaining bus rapid transit system', the system sustained itself on the basis of higher fares and funding from international loans which will effectively be repaid by national taxpayers. These loans were used to pay for the infrastructural work, thus acting as a hidden subsidy to the private companies operating BRT (Gilbert 2008: 450–9).

In sum, a cursory look at the record of BRT in Bogotá shows the tensions of this 'success story'. This is in line with most public–private partnerships, which are promoted as being self-evidently beneficial, but more often than not entail an uneasy relationship between the public sector (which provides the funding) and the private sector (which provides the services on long-term leases). What, then, are PPPs and from where do they originate? A PPP is a contract between a private company and a government. Under this contract, a

---

[11] The inflationary impact of BRTs on transport fares has been observed in many other cities, with costs as high as US$1.05 per trip (São Paolo, Brazil). Most systems with a fare below US$0.40 are reported to be under financial stress.

private company 'finances, builds and operates some element of a public service' in return for payments from the public authority and/or from users of the service (in which case the private company therefore obtains a concession) (Hall 2014: 5). PPPs originated in the UK during Thatcher's government as an 'accounting trick' to bypass austerity and the limitations that it imposed on government borrowing and spending. In a similar vein, they have been heavily promoted in developing countries since the 1990s, as a way to bypass the fiscal austerity to which governments were compelled by international financial institutions with the adoption of structural adjustment programmes in the early 1980s. PPPs constitute a very important business opportunity for the banks, construction companies, and service operators involved as they secure long-term revenue, typically between twenty-five and forty years, for each contract. For this reason, corporate interests are heavily involved in the 'global marketing network' that promotes PPPs around the world.

The economic case for PPPs rests on the claim that they bring both macroeconomic and microeconomic benefits to the economy (see Loxley 2013: 489, for a critical review). At the macroeconomic level, the argument goes that PPP projects 'bring extra money', as the private sector finances help to address the shortage of funds faced by governments. Another claim is that infrastructural work realized through PPPs is associated with more transparency and less corruption than when such work is carried out by governments. At the microeconomic level, advocates of PPPs argue that the construction and management of projects benefit from the superior efficiency of the private sector vis-à-vis the public sector, and from the 'capital market discipline' imposed on them by its financing from the private sector. This superior efficiency explains both cost savings and time savings in PPPs, which their proponents claim to be their distinctive features. Finally, the transfer of financial risk, from the public to the private sector, at every stage of PPPs' project cycle, is argued as a justification for PPPs.

However, the arguments made to justify PPPs are dubious, as closer scrutiny reveals. PPPs do not actually relieve governments from debts; they 'are simply debt in another form', that of the leases that governments agree to pay to private companies operating the services. These are more often than not more expensive than direct borrowing by governments themselves. Furthermore, the fact that funds are pledged for specific PPP projects has the effect of distorting policy priorities, as policymakers prioritize initiatives for which there is funding, and which are deemed commercially attractive. The claim that PPPs are less prone to corruption is contradicted by a wealth of scandals associated with them. In Tanzania itself, a PPP in electricity was defined by Transparency International as PPP 'at its worst' due to electricity being overpriced, not needed in the area, and not procured through a regular tender (World Bank 2012, Farlam 2005: 27, and Cooksey 2002; all quoted in Loxley

2013: 490). Experience with PPPs in the water sector has been similarly disappointing. Two years into a PPP that aimed to revamp the failed system of water supply in Dar es Salaam, the performance of the new consortium operating the system—called City Water, and led by British Bywater—was so poor and fraught with tensions that in May 2005, the Minister of Water unilaterally terminated its contract. City Water resisted the decision and as a result, following a police raid on the consortium office, its Chief Executive Officer was deported with two colleagues as 'undesirable immigrants' (Cooksey and de Waal 2007). In sum, the claim that the private sector, simply by virtue of being private, is more efficient and less prone to corruption than the public sector, is pure ideology. Furthermore, the claim that PPPs are time-saving is rarely born out, as PPPs, often involving very large-scale infrastructural work, tend to experience delays in their implementation. Private companies either protect themselves from this possibility with inflated bids or renegotiate contracts while the project is ongoing. An additional point against PPP efficiency claims is that competition between private companies for PPP bids is hard to find, as the number of eligible bidders for very large contracts is small. The relationship between such large companies and host governments is also problematic as it rests on significant information asymmetry. The former have a superior knowledge and experience of operating the service and exploit this to their advantage when negotiating the terms and conditions of a PPP. Finally, the claim that PPPs transfer risks from the public to the private sector sits at odds with the common practice, by bidders, of protecting themselves from a lack of demand with revenue guarantees. Furthermore, renegotiations of PPP contracts by and large favour the interests and revenue of the private sector (Trebilcock and Rosenstock 2015: 11).

In sum, the case for PPPs rests on very unconvincing grounds. What we know is that PPPs bring about 'the encroachment of private capital on public sector activities' (Loxley 2013: 494). More often than not, the superior quality of service delivery promised by its advocates does not materialize. Instead, what results are long-term, and tax-payer subsidized, lucrative leases to the private sector. As PPPs, BRTs are better understood as a solution to tackle the crisis in public urban transport in which international finance and the corporate sector have a powerful interest, as BRTs allow them to capture new markets and public funding (Paget-Seekins 2015). Their implementation has complex and often contradictory impacts. While BRTs are capable of delivering improvements in the standards of public transport, they also generate a set of tensions that typically coalesce around the lack of affordable fares, the exclusion of the previous transport investors and workforce in the new system, and the contraction of employment opportunities. This chapter now proceeds to analyse the implementation of a BRT in Dar es Salaam, in light of these tensions.

7.3 The Ideology of BRT in Dar: Whose 'Better City for Better Times'?

First announced in 2002, DART has been funded through a US$150 million loan from the World Bank to the government of Tanzania. In 2016, it was the most grandiose BRT to be launched in Africa, with the rebuilding and doubling in width of the main arteries in the city, for a total of 137 km of new road network, eighteen terminals, and 228 stations. The idea to implement BRT in Dar emerged as a suggestion by the Institute for Transportation and Development Policy (ITDP) to the City Council. ITDP then sponsored an investigative trip by the City Council to Bogotá (Kanyama et al. 2004: 43) and the delegation travelled back via an ITDP-facilitated meeting at the World Bank, at which the commitment of Dar es Salaam leaders to BRT was probed. This was clearly a success, as Kleist Sykes, the then Mayor of Dar es Salaam, was subsequently informed by the World Bank that Dar es Salaam had become the forerunner of BRT in Africa, allegedly overtaking Accra in the process (interview with Sykes, 2011).

Before analysing the tensions associated with its implementation, it is important to critically assess the rationale of DART and, as a project which aimed to radically rethink the city's transport system, to determine whether it adequately addresses the poor state of existing services. To be sure, large-scale infrastructural work and the widening of key roads were a much-needed step to improve transport in the city. As we have seen, Dar es Salaam had grown rapidly since the late colonial period, while its road system had remained by and large unchanged since colonial times. In addition, the number of vehicles had increased exponentially (Briggs and Mwamfupe 2000). Action was also needed to improve the poor conditions of public transport in the city, which were provided by primarily old, overloaded, and unsafe *daladala*. However, it is important to ask whether there was no alternative to the phasing out of *daladala* and to question the rationale for providing two exclusive lanes to BRT buses.[12]

The 2009 National Road Safety Policy put partial responsibility for the inefficiency of the transport system on 'rapid increased car ownership' (United Republic of Tanzania 2009: 36). Indeed, DART itself acknowledged this when it stated in 2014 that in Dar es Salaam there are '120,000 private vehicles that carry only six per cent of residents with 480,000 of their seats lacking passengers' (DART 2014). DART's vision of a better transport system and its desire to avoid any clash with private cars owners, the rising affluent, was a political one—one which leaves the inefficient use of private vehicles, by far the largest

---

[12] According to the DART plan, 150-seater buses will serve the main roads, while on feeder roads smaller buses, carrying approximately fifty passengers, will operate. This increased carrying capacity of buses, from the current scenario in which the majority of vehicles are thirty-five-seater buses, is aimed at easing traffic congestion.

cause of traffic congestion, unchallenged. At the same time, DART proposed that 'the current state of affairs where Dar es Salaam has more than 6,000 commuter buses that carry *only* (emphasis added) 43 per cent of city dwellers is not sustainable'. Equally problematic is that *daladala* were identified as the sole culprits of transport problems and were therefore not to be permitted on the two lanes exclusively set aside for public transport. It is worth reflecting on the fact that the types of buses to be operated by DART were taken as a given, and no thought went into supporting the improvement of the *daladala* fleet with, for example, effective regulation and a recapitalization programme to scale up the average size of buses and reduce their average age. Therefore, the rationale of DART and its prioritization of the solutions to public transport chaos in Dar es Salaam were questionable from the outset.

Finally, there was the thorny issue of fares. A 2009 survey of 150 public transport users found that 78 per cent of the sample deemed current fares charged by *daladala* expensive. Whilst at that point DART had yet to announce the official level, a further fare increase, in a context in which the mean expenditure on travel was as high as 24 per cent of passengers' monthly individual salaries, was likely to generate tensions and force some people out of motorized public transport (Ahferom 2009: 20). These issues were not peculiar to DART, but rather general trademarks of BRTs.

7.4 Making Sense of Delays in the Implementation of DART

What is peculiar to DART is the slow pace at which implementation has progressed, resulting in the loss of its status as BRT forerunner in Africa. Cape Town and Johannesburg, for example, embarked on BRTs well after Dar es Salaam but completed their projects earlier—although not without similar tensions (Barrett 2010; Schakelamp and Beherens 2010, 2013). While DART officers have exclusively attributed the slow implementation to the size of the project (interview with Schelling, 2014), there is arguably more to it. The following analysis focuses on two layers of explanation behind the delay. The first includes the practical barriers against which DART stumbled: that is, the concrete sites around which the tensions between project implementers and affected parties coalesced. But there is a second layer, that of the deeper roots of the delays in implementation, to be explored.

The obstacles faced by DART implementers changed over the years, but two disputes stand out due to their prolonged nature. These were concentrated around two sites of key importance to the project. In both cases, resistance was collectively orchestrated and changes in the institutional set-up of DART triggered the disputes. The project was initially owned by the City Council. Shortly after its launch, however, it was felt that the project would proceed slowly if it remained under City Council leadership, and that, given its national importance, it required highly qualified personnel. DART was thus established as a unit under the Prime Minister's Office (interview with Sykes, 2011). The Prime Minister at that time was Edward Lowassa. He replaced Kleist Sykes, DART's champion and Acting Chief Executive Officer, with Cosmas Takule. A number of informants agreed that Takule, an accountant by trade, had no skills to lead DART, and that his selection was motivated by Lowassa needing 'one of his men' for the post of DART CEO, to control the funds associated with its implementation.[13] While no major corruption scandal emerged around DART's implementation under Takule's leadership, the involvement of Lowassa in other corruption cases made such claims plausible. Furthermore, as I will show, the inadequacy of DART's new chief executive was evident in his poor leadership and his failure to negotiate the support of the local authority to facilitate the implementation of the project.[14]

The first obstacle to implementation was over the use of Ubungo station for which the Council and DART had conflicting agendas. Ubungo station is Tanzania's largest terminal to upcountry destinations. As the funds that the government pledged to transfer each year to the City Council rarely materialized in full, revenue from Ubungo was an important asset to the City's policymakers.[15] However, Ubungo station was also key to DART. Ubungo was planned as one of DART's five main terminals and one of its two bus depots, using 52 per cent of the existing terminal area. The proposed change of land use posed no threat to the Council while it administered DART. Having lost ownership of DART, however, the Council faced a future loss of revenue and no control over the new planned use of the Ubungo terminal area. These circumstances explain the hostile remarks of the Ubungo Station Manager, who complained that 'what is wanted is for them to find another place rather than taking an area which already has a specific and important function' (interview with Ubungo Station Manager, 2011). The Council's objections had some purchase, as a revised plan for the construction of a temporary upcountry terminal ensured that in making space for DART at Ubungo, the City Council at least secured new land nearby. However, the Tanzanian President intervened, condemning the move as 'a misuse of public funds' (*Daily News*, 20 September 2012). In blocking the initiative, he sent the two parties back to the drawing board to find a permanent rather than a short-term solution. Two months later, Takule lost his job. Local press suggested that this happened due to the 'president's doubts about [Takule's] request of

---

[13] The three informants who made this claim asked to remain anonymous.

[14] The same three informants converge in underlining these weaknesses in Cosmas Takule's leadership.

[15] The Ubungo Station Manager claimed that revenue from the station was worth 50 per cent of the council's direct revenue.

2 billion Tanzanian shillings to build a temporary terminal' (*Raia Mwema*, 21 November 2012). It is also significant, however, that Edward Lowassa was no longer Prime Minister by this point, and thus Takule had lost his political protector.

The second area of conflict involving DART was over compensation, not for lack of funding but rather as a result of the lack of support by the Tanzanian institutions executing it (Ka'bange et al. 2014). The Tanzanian President blamed DART management's ineptitude, when he 'blasted [its] Chief executive . . . whom he accused of stalling the project by failing to pay compensation on time to people who are supposed to be relocated' (*Raia Mwema*, 21 November 2012). However, lack of clarity over which Tanzanian public institution had to support DART in the compensation process was also to blame. Thus, in January 2011, Magufuli, then the Tanzanian Minister of Works and from 2015 the country's President, 'warned that if the city authority will not take immediate measures to pay compensations, his ministry will start to demolish places where the project is expected to pass' (*Daily News*, 16 January 2011). A few days later, the City Council Director responded that 'it [was] the government's responsibility to pay compensation—not the city authority' and that the City Council was 'also longing for the project to take place'.

Notwithstanding this confusion over the roles to be played by different state institutions, the responsibility for evaluating properties to be compensated undoubtedly lay with the City Council. The Council's Principal Valuer commented that the Council 'will no longer continue with the exercise due to other commitments as communicated to DART agency' (Chidaga 2010: 3–4), which suggests that the Council undertook this role with only tepid commitment. Taken together, the slow pace at which evaluations were undertaken and the lack of clarity over which institution had to pay compensation meant that considerable time passed between the valuation and the payment of compensation, resulting in significant numbers of claims for revaluation due to inflation (DART 2010: 24–249). Furthermore, managing the compensation of business tenants for their loss of profit was complicated because most tenants renting from the Council had sub-leased. The Council considered that they had 'no legal contracts with the city council, thus [they] do not qualify for compensation' (Chidaga 2010: 3). John Mnyika, the Ubungo Member of Parliament for CHADEMA, the leading opposition party in Tanzania since 2010, disagreed and supported the resistance of informal (and formal) tenants to resettlement until a place of comparable value for their business could be found.[16] These multiple turf wars at Ubungo station, some resolved as late as 2014, long delayed the DART project.

---

[16] Anonymous informant, but see also several entries on Ubungo station in John Mnyika's personal blog, n.d. <http://mnyika.blogspot.co.uk/>, accessed 10 November 2014. On opposition parties in African urban settings, see Resnik (2012) and Lambright (2014).

The second main site of resistance to the project was Gerezani, in the Kariakoo area of the city centre, where DART planned to locate a terminal, its main office, and its control tower. Within this area there are fifty-three twin houses with a total of 106 owners. Twenty-nine owners accepted the proposed compensation for demolition, while the rest brought a court injunction against it. However, the court judged that the occupiers of these Gerezani houses had no right of occupancy and instructed their demolition (United Republic of Tanzania 2008). Taking advantage of delays with the implementation, in itself a sign of the limited urgency given to the project by public authorities, the occupants registered themselves as the Kibasila Society Group and made a further court appeal. Fuelling this prolonged battle was the fact that DART had sidelined local councillors and overlooked 'the importance of disseminating correct information about the project' (DART 2010: 4). As Mr Natty, Acting Director of Dar es Salaam City Council in 2011, put it, 'If they had used local councillors to mediate with these residents it would have been easier than using DART technical experts. You know communities have a lot of respect for their leaders' (interview with Natty, 2011). In addition, the fact that 'PAP (Project Affected People) are given contradictory information like the project has been abandoned' (DART 2010: 4) suggests that those who intended to boycott DART exploited its public relations failure. The Gerezani case contributed to the delays in the project and was resolved only as late as 2012, when the Tanzanian court settled the issue by setting the value of the compensation and by allowing DART to demolish the houses.[17]

These contestations, which highlight the local politics of resistance to DART, suggest that the project was perceived as a threat by some, despite its own 'win–win' rhetoric. In summary, those who risked being made worse off, including the City Council which faced a loss of revenue, tenants who had to be relocated, and individuals threatened with expropriation of their property, resisted the project by leveraging a number of strategies, such as withholding or slowing down the public support that the project needed to make progress, legally challenging the process of expropriation and compensation, and drawing on an opposition MP who championed the grievances of the citizens affected by the project. In so doing, they exploited the ineffectiveness of DART's leadership.

An aspect that stands out when scrutinizing the main hindrances to the implementation of DART is the mismatch between the relatively small nature of the problems that caused the delay and the time that it took to overcome them. Given the strategic importance of DART, it is puzzling that seventy-seven

---

[17] The court case was no. 44 of 2012. See Kilasa Mtambalike, 'Dart compensation saga settled', *Daily News*, 31 May 2012 <http://allafrica.com/stories/201205310995.html>, accessed 31 January 2017.

house owners slowed the project by five years and that the compensation process proceeded at such a slow pace. Could no institution above the Ministry of Works and the Council step in to clarify their roles? And were there no other ways to solve the quarrel with the resisting house owners so that it did not result in years of standstill? Arguably, a lack of high-level commitment to the project by the Tanzanian government underpins such an impasse, the reasons for which are explored in section 7.5.

7.5 The Deeper Roots of Lack of Government Support

DART advocates promoted the project as a solution to urban transport problems that was beneficial to society and the economy at large, but as implementation neared, some groups stood to be negatively affected. A document entitled '*Daladala* grievances plan', written in early 2010 by a consultant to the project (Kamukala 2010), provides clues as to the tensions between the real politics of DART and its rosy portrayal, and to the often contradictory way in which these were handled. The document outlined how to manage 'the grievance process with *daladala* bus owners and operators in order to bring about a *win–win* (emphasis added) situation for everyone involved' (Kamukala 2010: 1). However, such optimism sat alongside awareness of how delicate the issue of incorporation of previous operators was: 'DART has to clearly inform the *daladala* owners and operators how the introduction of the BRT is likely to affect their overall employment opportunities in the public transport sector, otherwise there might be strikes and violence due to lack of appreciation of the BRT system'. Along the same lines, one reads that 'it is important for DART to let stakeholders understand that introduction of the BRT system is an opportunity and not a threat to their existence'. However, no evidence was given for why this might be the case.

There were several different reasons for resistance to the project by affected parties, other than the 'lack of appreciation' of BRT. The unknown yet likely negative impact of DART on employment and on the estimated 20,000–30,000 workers in the sector was problematic for the Tanzanian leadership and one of the two key causes for its half-hearted support. At a dialogue on unemployment organized by Japan International Cooperation Agency in July 2013, the then Tanzanian President, Kikwete, expressed his concern about the gravity of the problem of under- and unemployment by identifying the decrease in the rate of unemployment amongst the youth in Africa as central to future political stability. As he put it, 'The Arab springs were about government that overstayed but the unemployment spring is coming to Africa and will not spare democratic governments' (*Daily News*, 4 June 2013). The President also openly voiced his worry about the employment implications of

DART. At a show to publicize the project in Dar es Salaam in 2006, Kikwete stated that DART was going 'to reduce employment' (interview with Sykes, 2011). Kleist Sykes, Mayor of Dar es Salaam from 2001 to 2005 and a major force in the project since 2002, protested that this fear of unemployment was misconstrued as DART would 'create rather than reduce employment', thanks to training programmes for all affected parties, including workers, station attendants, and mechanics.

Yet, as I have shown, there was ground for scepticism about this smooth integration of pre-existing operators into DART predicted by its promoters. As 'a single DART bus will displace about 10 minibuses', job destruction will inevitably be considerable (*All Africa*, 24 August 2012). Although there were funds earmarked to train some drivers to qualify for driving on BRT buses, DART operators could not be under any obligation to hire them (interview with Schelling, 2011). In sum, DART entailed a shift from labour-intensive to capital-intensive urban bus transport, with a contraction in the quantity of jobs and, ideally, their replacement with more secure and better-quality jobs. The role of the sector as a source of employment for the urban poor was a goal of secondary importance to increasing the carrying capacity of buses. From this perspective, one can understand the dismissive tone with which the 'grievance management planner' addressed workers: 'Some of these drivers will be re-routed; absorbed in the BRT system; lose jobs etc. *They should categorically be informed what would be their fate*' (Kamukala 2010). The government, and first and foremost the President, seemed less enamoured with such a stance.

Difficulties in incorporating *daladala* owners in DART constituted the second main reason behind the lack of support by Tanzanian leaders for the project. As previously discussed, the involvement of pre-existing transport providers in BRT often proved to be a thorny issue. Dar es Salaam is no exception. Such difficulties stemmed from the place of *daladala* owners in DART being an afterthought to its planners. A study in 2009 found that, although there was a 'strong will to involve current *daladala* operators', there was 'no clear plan on how that will be accomplished' (Ahferom 2009: 21). In May 2014, weeks away from the actual tendering process, lack of clarity persisted, as evidenced in the DART Project Information Memorandum, and in particular in its '*daladala* transition policy' (Rebel Group 2014: 38). One of its pillars was the resettlement of *daladala* to routes on the outskirts of the city. The Memorandum conceded that 'some *daladala* owners may then decide to withdraw from the public transport business altogether and invest in other sectors' but concluded that 'the fast population growth and rapid spreading of suburban neighbourhoods will create sufficient need to absorb most of this freed capacity' (Rebel Group 2014: 38). However, as DART aimed to capture more and more of the market share controlled by *daladala*, thus pushing them out of their areas of operation, the expectation that supply would create its own demand and that their business could be sustained when displaced to feeder routes was implausible.

The real issue was how *daladala* owners could become part of DART proper. Their inclusion was complicated because the bidding document for tenders for the two companies that would operate DART buses required that bidders must have been successfully engaged in something similar (interview with Schelling, 2011). As there were no Tanzanians with experience of this kind, this ruled out domestic investors from fully owning DART and fuelled tensions between foreign and national ownership of BRT implementation in Dar es Salaam. Indicative of the deep-seated tensions that characterized Tanzanian leaders' thinking on urban transport, Tambo Mhina, former secretary of the Dar es Salaam Region Transport Licensing Authority (DRTLA), recalls how in 2003, when the City Council had already endorsed the BRT proposal, his office prepared a plan to consolidate ownership of (larger) buses in Dar into two or three companies (see section 2.5). Such a plan did not convince high-level decision makers, who objected: 'These people who own one or two buses, they are original Tanzanians. If you advertise a tender for large buses and large companies, investors of international calibre will come. They [Tanzanians] have already invested in the sector and will be sidelined. Where will they put their capital?' (interview with Mhina, 2010). Therefore, as late as 2003, the year in which the Council formally endorsed DART and its planning began, the Tanzanian leadership clearly prioritized the interests of indigenous investors over and above foreign investors and considerations on how to improve urban transport. With the attempt to roll out DART, ten years later, the problem had presented itself again, and more pressingly so. How then were the interests of Tanzanian investors being protected this time?

At the official inauguration of the construction of DART in 2010, the President Jakaya Kikwete urged *daladala* owners not to 'just remain idle complaining about the arrival of foreign investors' (Stewart 2014). By this, he intended that owners should ensure that they were able to meet the requirements to operate the smaller buses on BRT feeder routes, and should form collective enterprises to join the operations on BRT main routes (as opposed to its feeder routes). Some owners had indeed been preparing for this, most notably the Dar es Salaam Commuter Buses Owners' Association (DARCO-BOA), which registered a company to this end (Mutasingwa n.d.).[18] However, a major barrier to joining BRT was the cost of buses and the lack of finance. Owners organized lobbying of various ministries, including the Ministry of Transport and Ministry of Finance, to facilitate support in the form of a loan

---

[18] The company is called Cordial Transportation Services (PLC).

guarantee and tax exemption, yielding promises but no tangible results. DAR-COBOA responded by setting up a meeting with 'the Member of Parliament for Dar es Salaam and Councillors who intend[ed] to be involved in the 2010 country election as candidates, to help them to sell the idea', as they felt the issue required 'political will'. The government offered public funds to promote *daladala* owners' incorporation into DART in September 2012, when it announced that it would purchase all old *daladala* six months before DART was due to start operations (*Daily News*, 25 September 2012). Thus, similar to the cases of Johannesburg and Cape Town, political pressure resulted in public funds becoming available to owners for the purchase of 'free' shares of the new BRT (City of Johannesburg Metropolitan Municipality 2011; Schakelamp 2011: 10–11). However, the bus owners' association turned down the offer, as the compensation that could reasonably be offered for fairly old vehicles was expected to be low (interview with Mabrouk, 2014).[19] Furthermore, many bus owners doubted the possibility of making a profit on shares without being cheated. Instead, owners restated their request for government support in the form of its guarantee to access finance for purchasing new buses. But as the chairman of the bus owners' association explained, the failure to achieve any breakthrough had to do with reasons beyond government control: 'In the words of the World Bank, there is no way we would give a loan guarantee to the private sector . . . I told people at the World Bank, that if we continue like this, it means that you want us [*daladala* owners] to get out. To us this is an elimination process' (interview with Mabrouk, 2014).

Arguably, the World Bank and DART were merely paying lip service to the importance of local ownership of the project, since, according to the DART Chief Technical Adviser, a retired World Bank Transport specialist, 'whatever we do with *daladala* has to be compatible with the market. We cannot sell the project unless this is the case' (interview with Schelling, 2014). The implication was, in terms of the policy priorities on DART in 2014, that the foreign companies that successfully bid to operate DART, and from which *daladala* would have to subcontract their business, will have a key role in determining the terms and conditions of their service. What remained unexplained was the way in which Tanzanian bus owners were expected to compete in the market with much larger and wealthier foreign transport companies without any external support. In the words of DARCOBOA's Chairman: 'Even in boxing, Prince Nasseem [a featherweight world champion], you cannot put him with Mike Tyson [heavyweight former world champion]. He will kill him, you cannot do that. This should be the logic, people with the same weight compete' (interview with Mabrouk, 2014). Along the same lines, Engineering

---

[19] The value of the buses became a hotly contested issue in other cities. On Bogotá, see Gilbert (2008).

Professor Mfinanga, a researcher on urban transport and on the Board of Directors of DART, 'cautioned that expecting [owners] to form companies and compete in an international bidding contest is "insulting to them"' (Stewart 2014). This is the way things stood, unsolved, as tendering became imminent. However, one would not remotely sense this from the Project Information Memorandum's rosier version of the situation: '[Bus owners] are generally supportive of the DART initiative . . . This industry restructuring and reorganisation process is also meant to give *daladala* owners a chance to associate with local and international parties in bidding for the DART system and participate in the tender process' (Rebel Group 2014).

While the inconclusive search for ways to incorporate previous bus owners unfolded, DART faced a new challenge in the rapid growth of *Shirika la Usafiri Dar es Salaam* (UDA), a Tanzanian company, which sought to claim a key role in the project. UDA was the state-owned company that provided public transport in Dar es Salaam under a monopoly regime from 1970 until 1983, when private operators were permitted into the sector to satisfy the demand for transport that UDA could not match (see sections 2.2 and 2.3). As the decrease of its fleet progressed over the years and buyers could not be found when it was initially offered for privatization, UDA laid moribund, operating an average of as few as seventeen buses a year from 2002 to 2008 (National Institute of Transport 2010). However, the scenario for UDA has looked much brighter since the early 2010s. In February 2014, the company owned just under 400 buses and had ordered a further 1000 buses, which when delivered would make it the largest urban passenger transporter in Africa. Its ambitious plan was to field 3000 buses by the end of 2014 (*The Guardian*, 17 February 2014; *Daily News*, 11 February 2014).

Such dynamism was a result of UDA's privatization in 2011 at a price and through a process that generated controversy and a legal case (*The Guardian*, 29 January 2012).[20] UDA had been previously co-owned by the Dar es Salaam City Council (with 51 per cent of its shares) and by the Tanzanian government, holder of the remaining 49 per cent of the shares. By purchasing the City Council's shares, Simon Group became the largest shareholder of UDA, and Robert Kisena, the Executive Chairman of Simon Group, became its new CEO. Kisena was purportedly closely connected to the son of the then President of Tanzania, Jakaya Kikwete, who was believed to be the main man

---

[20] Prominent CCM leader and businessman Iddi Simba, then chairman of UDA board of directors, was involved in the controversy as he was charged, and then declared innocent, with forgery and abuse of office for selling UDA shares at a far too low price and for cashing in the money from the transaction. See Faustine Kapama, 'State Files Fresh Charges against Iddi Simba'. *Daily News*, 1 May 2013 <http://allafrica.com/stories/201305010352.html>, accessed 11 November 2014.

behind UDA's rise.[21] While there was no evidence to verify this claim, UDA's ambitious 'future plan [...] to be involved in DART' was clear (UDA leaflet n.d.). Unlike those of DART, UDA's public statements seemed more attentive to the employment dimension of its operations. As Kisena explained when commenting on the introduction of 300 new UDA buses, 'the company's move will also help in the fight against unemployment as the 300 buses, once operational, would provide 900 direct employments and about 2,000 people will be employed indirectly' (*Daily News*, 8 October 2012).

Although UDA's rapid growth and self-promotion as a DART operator were well-received by public institutions, it caused bitterness amongst *daladala* operators and DART (*The Guardian*, 15 February 2014). For the latter, a particular concern surrounds UDA's procurement of thousands of vehicles without consultation with DART. As DART's Chief Technical Adviser explained, 'UDA buses are not the buses we need. We cannot use them. Our stations are designed for specific buses, with bus stops and lanes designed to match them' (interview with Schelling, 2014). Therefore, the foundations for a new clash were laid between the technicalities of bus designs and project specifications and the 'made in Tanzania' version of BRT promoted by UDA. While the accusation of a connection between UDA's new owner and the son of the current Tanzanian President, Jakaya Kikwete, is unverifiable, one can see how UDA intended to seize a lucrative contract as one of the DART service providers by virtue of being the only large Tanzanian transport company that could supply buses to DART. As UDA directly addressed the anxieties of Tanzanian leaders over the impact of DART, most notably because it was owned by Tanzanians and aimed to employ a larger workforce, it was bound to continue enjoying the support that the Tanzanian authorities had afforded it since its revival.

7.6 Towards the Implementation and Domestication of BRT: 2014 Onwards

As the beginning of its operations neared, and so did the presidential elections of 2015, the political heat surrounding the project increased. The stakes in incorporating (or not) bus owners in the new system were very high, and the actual impact of DART, including its displacement effects, were to become

---

[21] See discussions in *Jamii Forums* <http://www.jamiiforums.com/jukwaa-la-siasa/159840-uda-connetion-ya-robert-kisena-na-msharika-wake-ridhiwani-kikwete-4.html>, accessed 12 November 2014. Three informants, whose identity cannot be revealed, concurred with this rumour. Robert Kisena was an MP candidate at the parliamentary elections for CCM in 2010, when he lost to an opposition candidate. See Florian Kaijage, 'UDA Scam: Simon's Damning Version', *The Guardian*, 29 January 2012.

tangible. The chairman of the bus owners' association stated in May 2014 that 'we (i.e. current bus owners and Tanzanians) need to run the system on our own' (*The Guardian*, 26 May 2014). But a meeting for prospective bidders a month later saw investors from seventeen countries attending, including from 'South Africa, Brazil, China, Sweden, USA and Mexico' (*Daily News*, 22 June 2014). Mr Mushi, speaking on behalf of Dar es Salaam's District Commissioner, urged Tanzanian bus owners to strengthen their newly formed company 'to be able to compete in tenders'. Meanwhile DART, through its website, called for 'Politics [to] be left aside and [to] leave experts to do their jobs' (DART 2014): surely an implausible plea, as DART was inherently political, and, as such, the battleground of different political constituencies who continued to clash to promote, influence, or resist the project.

Delay following delay, in April 2015, the deadline for the beginning of the interim operations arrived. With it, the contradiction between presenting DART as a win–win intervention and its actual impact became more obvious. The modalities of DART implementation reflected how planners negotiated competing interests around the project. Although 90 per cent of the construction was complete, there were no available buses for DART and, apparently, none had even been ordered. Furthermore, no operator had been issued a licence to run BRT (*The Citizen*, 16 April 2015). What was happening? 'Powerful interests delay rapid buses project' read the headline of a Tanzanian newspaper. DART, went the story, was 'contending with two forces: one preferring foreign operators and another rooting for local transporters'. The tensions in the relationship between the World Bank and the Prime Minister's office over the matter seemed to crystallize this:

'The World Bank, which is the lender in this project, is playing some tricks, which will eventually delay this project unnecessarily,' said a senior official in the Prime Minister's office, who spoke on condition of anonymity because he is not authorised to comment on Dart affairs, which are in the Prime Minister's docket. 'But there are also foreign investors with interests in this project who want to be included in the deal.'

While UDA was keeping a close eye on these developments, it was careful to make clear, through its CEO's public statements, its view on who had the authority to settle the issue: 'Let us wait for the tender as it's the government that has the final say on who should operate this project' (*The Citizen*, 17 April 2015).

A week later, on 24 April 2015, DART shelved the idea of tendering for BRT, and instead signed a contract, without any formal tendering, which gave the UDA-RT, a newly formed company, a two-year licence to provide interim services to DART. The will to accommodate domestic interests into DART underpinned this interim measure, as the origin and composition of

UDA-RT, as well as the modalities under which it was allowed to operate, reveal. UDA-RT was a company formed by UDA and by the *daladala* owners' associations. With the interim contract, both Tanzanian bus owners and UDA were therefore established as the provider of DART for at least two years, and most likely beyond. As DART CEO explained, as 'the current providers of public transport in Dar es Salaam, [they] deserve to be part of the future transport system, but they have no experience in operating a BRT system, nor with the planned Smart card based fare payment system' (*All Africa*, 10 May 2015). The interim contract therefore crucially allowed for domestic actors to earn two years of operational experience. More than that, as the CEO made clear, when time for a competitive bid came, the successful bidder was to show 'willingness to integrate with the Interim Service Provider'. This was a change of direction in planners' priorities, as in earlier years DART leadership had stressed that only 'the market' could determine whether domestic and/or international competitors could be allowed to provide DART services.

While the interim contract was a statement of clear intent to bring local investors into the project, it only involved seventy-six large buses, a number which clearly could not accommodate the vast majority of an estimated number of 1651 bus owners which stood to be displaced by the beginning of Phase 1 of DART operations. In a move that further signalled the Tanzanian authorities' attention to accommodate as many domestic interests as possible, and in a clear breach of the original plans of DART promoters, under the new scenario the interim providers were to supply their services in competition with *daladala*. UDA-RT buses were to operate on the two segregated lanes, in parallel to *daladala*. The latter, rather than being displaced from service by DART, were allowed to use the two lanes originally destined by DART to private cars, alongside them. The rationale for doing so was also to shield the public from the significant increase in fares that DART was bringing about, and protect DART from public outcry over fare increases. As its CEO announced, the new system gave choice to passengers, who could decide between faster and more expensive UDA-RT buses or cheaper and slower *daladala* ones. While successful at accommodating as many interests as possible in the short term, such a policy had a ring of short-termism to it, as it meant that DART contractors lost their status as exclusive providers on their routes and with it, a significant share of the market. The business model behind DART, and in particular the revenue of DART operators, which was based on the assumption of full market control, was therefore sacrificed to the need to minimize the displacement of pre-existing operators. While the problem of the decreased profitability of DART providers was unlikely to go away, this would not re-emerge until the beginning of DART operations. The imminence of the presidential elections, at that time only five months away, seems an important explanation as to why the government might have decided not to choose between DART and domestic investors at that particular point in time.

7.7 What Can President Magufuli Do?

However, deadline after deadline continued to pass, unmet, and operations had still not started in October 2015, when the election of President Magufuli hit the Tanzanian political scene like an earthquake, with important consequences for DART. Relatively unknown to the wider public, his selection as the CCM (the ruling party) candidate for the 2015 presidential elections was surprising to many. Magufuli was Minister of Works from 2010 to 2015, and his selection was partly due to his success and resolution in delivering public works, something that earned him the nickname 'the Bulldozer'. The fact that while in that position, which entailed the management of several multi-million contracts, Magufuli was not implicated in any case of corruption, also played a big role in his nomination, as the CCM saw in his record the potential to fence off one of the main arguments used by the opposition to win voters' votes: CCM's failure to tackle corruption.

Magufuli's rise in popularity with both the Tanzanian and international public since becoming President has been phenomenal. Such public euphoria followed a number of bold and highly symbolic steps that Magufuli has taken since his election, all part of his strong drive to tackle corruption and cut unnecessary expenditure by public-sector employees (Gaffey 2015). For example, 2015 was the first year in Tanzanian history in which Independence Day was not celebrated. Due to a cholera outbreak in parts of the country, the President declared that it would have been 'shameful' to throw money at wasteful celebrations. Instead, he called on every Tanzanian to pick up a brush, sweep the streets and clean them, himself leading by example. Another highly symbolic move was to ban foreign travel for all government representatives, except for himself, his Vice President, and the Prime Minister. All government officials were directed to 'instead go to the villages to learn, understand and solve a myriad of problems our people are facing'. Heads of various government agencies started rolling, as evidence of corruption and mismanagement was found during a number of unannounced inspections to public offices by the President himself. In one of them, a visit to Dar es Salaam's Muhimbili Hospital, Magufuli responded to the deplorable conditions in which the hospital was found, with patients forced to share beds or to sleep on the floor, by diverting 200 million shillings earmarked for 'parliamentary parties' to buy 300 hospital beds. Magufuli also slashed the cost of his presidential inauguration ceremony from US$100,000 to US$7,000, donating the difference to the hospital. The message that emanated from these various actions, his intention to take money away from politicians and give it to the needs of Tanzanians, reached Tanzania and beyond loud and clear. The hashtag '#whatwouldmagufulido' became 'the most talked-about hash tag on social media' of 2015 in East Africa (*The Monitor*, 13 December 2015). It was employed by Twitter users who, inspired by Magufuli, posted photos of endless and often hilarious ways in which money could be creatively saved by ordinary people.

Without belittling the importance and galvanizing effects of these symbolic moves, tougher challenges undoubtedly lie ahead for Magufuli. His calls to reduce poverty through industrialization and job creation for the youth, or to use revenue from natural resource extraction for the benefit of Tanzanians, while promising, will inevitably lead him to step on the toes of powerful interest groups, both foreign and Tanzanian, that benefit from the status quo. This raises the question whether Magufuli has enough power to make these changes happen. It is also worth noting that the targets of his symbolic steps in the early days of his presidency were actors whom he could reach relatively easily: heads of government agencies, members of his own government, and MPs. In January 2016, a new addition to this hit list introduced a worrying xenophobic ring to his strategy, as Magufuli suggested that illegal immigrants working in Tanzania should 'surrender jobs to the nationals' (*Afrika Reporter*, 10 January 2016). Kenyan teachers working in private schools were his main target at that point. One wonders whether Magufuli will deploy this 'anti-foreign' narrative to target other, more powerful, foreign interests with stakes in other sectors, including DART, to which the analysis returns.

How has 'the Bulldozer' dealt with DART's seemingly endless delays? One month after Magufuli's election, in October 2015, the government attempted to inject momentum into the project through a Prime Ministerial Order that called for 'a speedy implementation of DART', and set a new deadline for 10 January 2016 (*The Citizen*, 7 December 2015). However, more bottlenecks stood in the way of the beginning of DART operations. Reservations by a government agency about the trustworthiness of DART providers, with implications as to the fare level to be charged by DART, lay at the heart of these. This time the problem was a political standoff between UDA-RT and the Tanzania Revenue Authority (TRA). While the former had imported 140 buses, they remained idle in Dar es Salaam port, as the TRA was unwilling to release the buses until the import duty, worth US$3.8m, had been paid in full (*The Guardian on Sunday*, 20 December 2015). Sabri Mabruk, whom we met as chairman of the *daladala* owners' association earlier in the chapter, and who by 2016 had become UDART board chairman, complained that the sum was 'a huge amount which we can't possibly manage to raise in such a short period of time. We are not asking for tax relief but to settle the tax obligation in instalment[s] as we continue to provide service to the city commuters' (*The Guardian on Sunday*, 20 December 2015). However, such a request was not to be met, due to an uneasy precedent between the two institutions. When the UDA—as opposed to the UDA-RT—purchased hundreds of new buses, it similarly requested a delayed payment of tax to the TRA, which the latter seconded. However, as things turned out, TRA found that the company 'failed to honour their promise' and this explained their unwillingness to agree again to a delayed payment of taxes (*The Guardian on Sunday*, 20 December 2015). Furthermore, doubts were raised about the UDA-RT's capacity to repay the TRA, given that the company had another debt, incurred when these buses were purchased with a separate loan that also needed repaying. In dismissing these doubts, UDA-RT at the same time put pressure on the government: 'It is quite possible to repay both as we are sure the project will be profitable. But this will largely depend on the government's consideration of our proposed commuter fare, as it is *the only source of our revenue* (emphasis added)' (*The Guardian on Sunday*, 20 December 2015).

However, when the fares proposed by the UDA-RT became public, the problems became more entrenched rather than easing. While a *daladala* fare cost between 400 and 600 Tanzanian shillings, UDA-RT's proposed fares ranged from 700 to 1,400 Tanzanian shillings. Public outcry about what constituted more than a doubling in public transport fares unsurprisingly followed. 'UDA-RT are simply thieves' stated a Swahili newspaper (*MwanaHalisi*, 5 January 2016). The SUMATRA Consumer Consultative Council dismissed the fares as both unviable and inflated. It asked for evidence of the way in which the UDA-RT business plan justified the fares requested (*The Citizen*, 14 January 2016). The Prime Minister was equally unconvinced in his remarks that 'the proposed fees defeated the intention of initiating the project' (*The Citizen*, 11 January 2016). He went on to add that 'the proposed fares are too high. When the government built this project, it intended to help civil servants and other low-income city inhabitants reach the city centre in time and at affordable fares' (*The Citizen*, 14 January 2016). As the director of Twaweza, a Tanzanian NGO, underlined, the proposed fares would have meant 'that a common Tanzanian was going to spend his total daily income on transport only' (*The Citizen*, 6 January 2016). The spectre of nationalization was also authoritatively waved by the Prime Minister at UDA-RT, when he stated that unless it reduced the fare requested, 'the government would take over the operations of the project' (*The Citizen*, 14 January 2016).

Alongside the problems with the taxman and the impasse on fares, a corruption scandal hit DART, when its CEO, Asteria Mlambo, became the fifth head of a government agency or public corporation to be axed since Magufuli's election. In December 2015, the CEO was suspended over alleged irregularities with the procurement of the interim service agreement (*The Citizen*, 24 December 2015). The main allegation against her was failing to take action when UDA-RT violated their contract, signing an interim contract without competitive bidding and failing to duly involve the Board for key decisions on the project (*The Citizen*, 11 January 2016). Intriguingly, doing without a competitive tendering process was a possibility that the CEO had ruled out in earlier years, subsequently softening her position precisely to facilitate the inclusion of domestic investors into DART. As late as January 2014, Mlambo stated that DART operators had to be obtained 'through a competitive bidding in conformity with the laws of the land' (*The Citizen*, 27 January 2014). In May 2015, her position was that the selection of UDA-RT as an interim provider without competitive bidding was legitimized by the procurement laws, under 'the purpose of creating local capacity' (*Daily News*, 10 May 2015). Arguably, the political pressures for the inclusion of the domestic investors played an important role in this changing interpretation of the procurement law, and ultimately in the suspension of the CEO as the government sanctioned her choice.

The new CEO of DART, Ronald Lwatakare, had a hot potato in his hands (*Tanzania Invest*, 22 January 2016). The political standoff between the taxman and BRT's interim provider, the latest obstacle to the beginning of operations, had ricocheted into a public debate about the affordability of fares. What power did the government have to unlock the situation? The debates continued, including calls for a public subsidy for public transport, which UDA-RT suggested was the only option that allowed both affordable public transport and a profitable business for its provider (*The Citizen*, 6 January 2016). The government, while not dismissing the option *a priori*, identified understanding the irregularities of the project as the first step to finding a solution (*The Citizen*, 11 January 2016). However, another commentator questioned why, if DART operators could not provide cheaper or at least the same fares as *daladala*, they should be subsidized to provide a service on infrastructure built with public money. Instead, he demanded free competition between different actors (*The Citizen*, 21 January 2016).

There was no easy way to settle the impasse on fares, a precondition for the beginning of operations, as the content of the contract previously signed by the interim provider and DART constrained the government's room for manoeuvre. While the Prime Minister was seeking to redesign the operational terms, UDA-RT invoked compliance with the agreed ones. According to Mr Kisena, its Managing Director, 'The arrangement is that the project is protected against any form of interference. So, if today the government says it is not in support of DART, it will have to take over the loans or at least leave the operations in the interim to the operators alone' (*The Citizen*, 22 January 2016). Neither of the two possible scenarios was straightforward, and this explains why, despite the urgency of the matter, and three weeks of meetings led by the Prime Minister with various public transport stakeholders, things stood at a standstill in March 2016 (*The Citizen*, 2 March 2016).

In May 2016 DART finally made its operational debut. The fares actually charged by UDA-RT, ranging from 650 Tanzanian shillings for trunk routes to 800 for feeder and trunk routes, represented a significant increase from the 400–600 Tanzanian shillings charged by *daladala* (*The Citizen*, 10 May 2016).[22] However, at the same time, they were much lower than what UDA-RT had originally proposed (1,200 to 1,400 Tanzanian shillings), thus showing the way in which the government protected the interests of passengers over and above those of DART interim providers. What remained to be seen, given the lower fare levels, was the sustainability of the operations of DART providers, and whether the need for subsidizing them would creep back into the debate.

As well as keeping fares as low as possible, the government also stamped its authority on DART by reclaiming its ownership of 49 per cent of UDA shares—emerging victorious from a legal case in which it was challenged—and as a consequence, of UDA-RT, the interim provider of the project (*The Citizen*, 10 May 2016). The legal battle started in 2015, when UDA's board of directors, under instructions from the company new majority shareholders, Simon Group, attempted to sell to itself the government's 49 per cent of the shares in UDA. However, the Treasury Registrar later established that the sale of the government shares was illegal, as there was no evidence to suggest that the government ever authorized the sale of its shares, and it was the only authority with the right to do so. The legal battle was settled, perhaps not by coincidence, just days before DART began its operations. Galvanized by this victory, the government voiced its intention of being more actively involved in the running of UDA, of UDA-RT, and, by default, of DART itself. As the Treasury Registrar put it, control of such a large percentage of share meant that 'We will start by addressing all the wrongdoings in this project and effectively manage its operations' (*Daily News*, 30 April 2016). In sum, when DART finally launched, it seemed a PPP in which the role of the public sector was much more central than in the original plans put forward by DART promoters: the public sector was not only a regulator and overseer of private operators, but also directly involved in the ownership and running of the interim private service provider. The government's control over DART seemed to tighten further in October 2016, when Simon Group, the majority shareholder of UDA, was declared bankrupt due to its failure to repay a loan of just over US$2 million to the public lender, the Tanzanian Investment Bank. Although

---

[22] School pupils' fares were set at a flat rate of Sh 200 for all routes.

Simon Group's Executive Chairman declared that he was in a position to repay the outstanding debt and condemned the move as part of a nationalist 'government crackdown on formerly privatized state corporations', the Group's assets were confiscated and a bankruptcy court appointed a law firm to run the company (*The Citizen*, 19 October 2016).

7.8 BRT Tensions as 'Actually Existing Neoliberalism'

This chapter has argued that the narrative of BRTs, as a 'win–win' intervention to solve the public transport crisis in developing countries, obscures the many tensions associated with their implementation. Such a narrative stems from research sponsored by international finance, its NGO brokers, and BRT vehicle manufacturers, and is functional to their interests in opening up public transport markets in developing countries. While public transport in most African cities is experiencing a serious crisis, BRT is a solution that must be understood as one of the latest—and most rapidly expanding—fronts of the promotion of neoliberalism in Africa through PPPs in urban public transport. By debunking BRT rhetoric, this case study of the BRT in Dar es Salaam helps to highlight the real politics and the main tensions and contradictions associated with one instance of 'actually existing neoliberalism'. More specifically, the chapter highlights the lack of any serious attempts to include the current public transport workforce by DART, and documents the destruction of employment opportunities and tensions over the incorporation of previous owners, compensation, and the affordability of the new service.

I have argued that such tensions have been problematic for the Tanzanian government and thus go a long way to explaining its unwillingness to resolve the conflicts that DART generated. Above all, the slow progress of DART stems from the tepid commitment to the project by the Tanzanian government, which reflects its attempt to bring into harmony the conflicting interests of the World Bank and the demands of a number of local actors to whom it is electorally accountable. In sum, the form that DART has taken was significantly different from what its advocates had in mind, and this reflected the capacity of a number of Tanzanian actors to resist being taken for a ride.

8

Conclusion

Taken for a Ride

8.1 Cities of Ghosts: Bringing People Back In

This book has been a journey through the history, economic organization, and politics of public transport in Dar es Salaam, Tanzania's largest city. It has been a journey from the state provision of the service (in the 1960s and 1970s) to its privatization, liberalization, and informalization, from 1983 to the present. Who owns what, who does what with it, who derives power from what, and how such power is exercised have been the central questions that have guided my efforts to make sense of the transport system, in all its messiness and through over thirty years of change. They bind together the book's shifting focus on different facets of who holds the key to power and precarity in the transport system.

The journey has made six stops. At the first, the shift from public to private provision of the service was explored. I reviewed the trends that influenced the demand and supply of public transport in Dar es Salaam from independence to 2016 and the failure, by the Tanzanian state, to match growing demand for public transport. The private sector entered into service provision in 1983, and by the early 1990s, *daladala* became de facto the only supplier of public transport. Such a momentous shift in service provision mirrors the effects of the rise of neoliberalism on policymaking, and the rolling back of the state brought about by the adoption of structural adjustment policies. The performance of the private sector, following the progressive deregulation of the service, reveals how urban passengers of public transport in Dar es Salaam have been taken for a ride literally and in a more figurative sense. The rationale behind the reform, which centred on the belief in the superior efficiency of the individual in 'free markets', proved to be an article of faith. While the supply of transport increased, the provision of public transport by the private sector was chaotic. This chaos triggered state intervention and, since the very late 1990s, when the process of liberalization and deregulation reached its peak, a number of initiatives to regain some form of public control have been attempted. These regulatory efforts signalled the Tanzanian public authorities' intention to reclaim some policy space, and their inability to have any major impact on public transport problems. Such were the dynamics of 'actually existing neoliberalism' (Brenner and Theodore 2002) that explain the relationship between the public and the private sector from 1983 to 2016 and its impact on public transport in Dar es Salaam.

At the second stop, I focused on the private sector. Asking who owns the buses and who does what in the provision of public transport threw light on a key aspect of its economic organization: labour relations. The chapter debunked a key tenet of neoliberal ideology on economic informality, and one that has a strong hold on policymakers, namely that entrepreneurial self-employment is the dominant mode of employment in the informal economy. By contrast, the chapter documented a clear distinction between bus owners and workers, whose only asset is their own labour. The balance of power between the two classes was analysed, showing how owners took advantage of the oversupply of unskilled labourers, by imposing upon them exploitative informal employment relations, which resulted in harsh working conditions and meagre and volatile returns. Economic exploitation of workers had knock-on effects on passengers, as it was workers' reactions to the lack of regulation of the employment relationship that explained the many inefficiencies and tensions which characterized the provision of public transport by the private sector. The chapter also showed how economic injustice goes hand in hand with statistical invisibility: the types of workers at the centre of the book hardly exist as far as official statistics suggest. A critical review of the origin of key terms such as 'paid employment' and 'self-employment', and of their misleading translation into Swahili, exposed the shaky foundations on which such statistical information rests.

Alighting at the third stop, it was noted that while unregulated labour relations were central to the way *daladala* worked, they were ideologically hidden from view by a discourse, relayed by the state and the public, which attributed the cause of public transport problems to the workforce. The analysis therefore investigated the causes of workers' failure to respond to this state of affairs (through an institution representing their interests and side of the story) and, more broadly, the sources of power by workers in the labour market. It also interrogated the significance of existing forms of 'inward-looking' solidarity among transport workers, emphasizing their role in *managing* the effects of precarious employment rather than *challenging* its causes.

The fourth stop was unscheduled and forced by events, as since the early 2000s, political quiescence had given way to political organization. The establishment of an association, founded by transport workers for transport workers, reversed the political asymmetry that had characterized urban transport policymaking until then. Over the years, in partnership with the Tanzanian transport union, the workers association had demanded employment contracts of employers and of the state. How workers mobilized power, and the strategy behind their more assertive and successful attitude towards the state and employers, became the focus. In the process, the analysis also engaged with ongoing debates on the possibilities of struggles for rights at work by informal workers.

The fifth stop was to take stock of the long-term dynamics of occupational mobility or immobility of *daladala* workers. Tracking the occupational whereabouts of 116 workers at two intervals, over eight and thirteen years, enabled investigation into whether or not work on *daladala* fuelled dynamics of micro-accumulation and upward mobility. Workers' occupational trajectories revealed the inadequacies of the mainstream narrative that sees work in the informal economy as a training ground where entrepreneurs earn the skills on which to draw in later employment. Upward mobility was limited and happened very slowly, if at all. Over half of the sample had not moved on in thirteen years, and ageing inexorably worked against the workforce. Among those who left, there were instances of exit as a result of marginalization as well as those which suggest upward micro-mobility.

The sixth stop was once more forced by events. The operational debut of DART, a large-scale public–private partnership designed to radically change public transport in the city, became imminent in 2015. The analysis therefore turned to the significance and political economy of DART in Dar es Salaam. The focus was on debunking the 'win–win' narrative on BRT by its advocates, and on identifying the vested interests behind its promotion. The chapter also reviewed how and why a range of Tanzanian actors resisted DART, thus delaying its implementation as they attempt to influence its shape.

In bringing our journey to a close, I tease out the broader implications of this case study of public transport and the possibilities of carefully theorizing from the streets of Dar es Salaam. I will first revisit dominant narratives on the African city and on economic informality, and their failure to read human agency and structure as mutually constituted, as has been attempted in this book. As neoliberalism was a key concept deployed in the book to understand the evolving nature of public transport in Dar es Salaam over time, the book will end by revisiting debates on the usefulness of the concept of neoliberalism in making sense of Dar es Salaam's transport system and the forces that are to be found operating within it.

This book started by underlining the shortcomings of influential contributions to the burgeoning literature on urbanization in developing countries and in particular in Africa. Despite their antithetical position, they share an unbalanced focus on either the structural forces that bring these cities into being, or the agency that the urban poor exert in making cities what they are. Apocalyptic narratives on the city score a crucial point when emphasizing the 'jobless' nature of much urban growth in developing countries and the poor state of urban services experienced by most of its inhabitants (Davis 2006). As fundamental is their attention to the failure of the postcolonial state's developmentalist project, and to the role of structural forces in causing the decline of the absorptive capacity of the formal economy while the number of job seekers has grown rapidly. Importantly, such a narrative exposes the shortcoming of mainstream fantasies on slum improvements and on the capacity for poverty eradication and economic mobility in the informal economy. However, for critics, such a narrative tends to overgeneralize and, above all, overlooks the agency of the urban poor in the city. It is therefore accused of depicting the urban poor as passive victims.[1]

By contrast, hope is the distinctive feature of much writing on the African city. A narrative on African cities, most influentially put forward by scholars such as Simone (2004a, 2004b, and 2005), Pieterse (2008), and Robinson (2006), has rejected as normative and Eurocentric readings of urbanization in the South as dystopic. It is misleading, they argue, to emphasize that which does not work in these cities. Instead, the trademark of this approach is its call to take stock of the agency of African urban residents in making cities what they are, as this allows appreciation of their functionality, rather than of their deviation from a Eurocentric notion of how cities should look.

While sympathetic to calls for attention to human agency, I have argued that the fundamental problem with this narrative is that it puts forward a shallow notion of hope, centred as it is on an understanding of the agency of 'ordinary' urban citizens, which is woefully silent about the structural constraints against which it must be conceptualized. Cities in such research are therefore extraordinary, as they appear, in a sense, as cities of ghosts. Their inhabitants float in mid-air, unhinged from the material and from economic structures. The implications of this shortcoming are both analytical and political, as romantic hope and unqualified celebrations of the poor's agency are no sound foundation for any attempt at radical imaginings and interventions in the African city. Instead, attention to what room for change there might be in a given place must rest on a more grounded understanding of the circumstances faced by the urban poor.

---

[1] Again, it should be noted that critics of Davis's apocalyptic tone forget that he planned a sequel book on 'the history and future of slum-based resistance to global capitalism' (Davis 2006: 201, 207). See Chapter 1, n. 3.

Contributions to the debate on the nature and growth potential of the informal economy similarly display an unbalanced focus on either structural forces or on the agency of the urban poor. Market fundamentalism lies behind de Soto's (1989, 2001) and many others' celebration of economic informality as a display of the agency by the informal poor against oppressive over-regulation by the public sector. The poor's agency manifests itself in their capacity to respond to market opportunities, and in their choice to engage in heroic entrepreneurialism to survive exclusion from the formal sector. However, such an approach conspicuously overlooks the role played by structural forces and by the lack of alternative employment in causing the forced entre-preneurialism towards which people at the bottom of the informal economy are pushed. It is a narrative that glosses over the linkages between the informal and formal economy, and focuses exclusively on self-employed entrepreneurs. Class and existence on insecure wage employment in the informal economy are therefore neglected.

A less rosy reading of the widespread informalization of economies con-versely emphasizes the role of class and of structural forces behind it. Atten-tion to the link between the formal and the informal sector, in stark contrast with the dualistic approach, led scholars such as Portes, Castells, and Benton to argue that informalization is better understood as a process, reflecting the strategy by formal businesses to avoid the costs of labour rights hard-won by organized labour (Portes et al. 1989). State alignment to the interests of capital and employers, rather than its over-regulation of the economy, is a necessary condition for the informalization of the economy. There is no room for hope in this narrative, as with the disappearance of the 'working class', the highly heterogeneous workforce that replaces it is depicted as powerless to resist an onslaught by employers. While there are important merits to this approach, a pitfall to be avoided is a functionalist reading of the growth of the informal economy entirely as a product of capital's and employers' interests, ignoring the manner in which labour is created and the ways in which it might resist capitalist attempts at domination.

Avoiding the shortcomings of 'all structure' and 'all agency' narratives on the city and on the informal economy, namely functionalism and shallow-ness, respectively, was one of the key goals that I set myself in writing this book on the origin, logic, and tensions of public transport provision in Dar es Salaam, as they developed over time. A less simplistic attention to the inter-play of structure and agency rested on rescuing, through a wide range of empirical data, the historicity and specificities of one context and the role of a range of international and local forces in shaping it. 'We are also humans' (see Figure 8.1), a *daladala* worker reminds us, and I hope to have enabled 'real, living' people to be brought back to the centre of the analysis: people who are often displaced from writing on the city and the informal economy through

Figure 8.1 *Na sisi watu. Tutaheshimiana tu* ('We are also humans. We will respect each other')

their portrayal as immaterial urban ghosts, as carriers of an inconclusively conceptualized alternative urban order, as free market champions or as passive recipients of overly abstract structural forces.

8.2 Grounding Neoliberalism

Grounding neoliberalism, and thus pinning down the interventions associated with it, has been key to understanding the concrete realities faced by workers and passengers of public transport in Dar es Salaam, and their influence on changes over time from 1983 to 2015. The findings of this book can therefore be seen as a contribution to the debate on the usefulness of neoliberalism as a concept to understand the world we live in and, more specifically, the urban experience.

To be sure, neoliberalism is a slippery concept, as it has been conceptualized in different ways by different people. In this book it has been understood as a political project, promoted by agents, and associated with a set of economic policies that, despite their wide-ranging focus, attempt to promote individuals' economic freedom and private capital as the best way to organize economies and societies. The ideological dimension of neoliberalism is also key, as its promotion rests on a certain way of conceptualizing how economies and societies work, with all their problems, which purposefully attempts to forge a consensus on what is to be done to address them. Furthermore, the policy agenda associated with neoliberalism has changed over the years, schematically, from the 'roll-back' neoliberalism of the 1980s to 'roll-out' neoliberalism since the 1990s (Brenner and Theodore 2002), and the way in which the project has taken root has been equally uneven. This is because the success of this political project, in individual places and sectors, rested on its agents' capacity to promote it and to negotiate possible or actual resistance to it.

For some, the multiple dimensions of neoliberalism, its changing focus, uneven reach, and manifestation in different places and times, call for the abandonment of the concept. Urban processes of change, Robinson and Parnell argue, 'are not easily attributable to something called neoliberalism' (Robinson and Parnell 2011: 528). Others, while not disputing that the dominance of the neoliberal project is never total, and that it takes 'variegated' forms in different places and times, conversely suggest that the rejection of the concept of neoliberalism implies 'an embrace of unprincipled variety and unstructured contingency' (Brenner et al. 2010: 203) and a loss of understanding of 'the underlying *commonalities*' shared by urban experiences across places.

The debate on the usefulness of neoliberalism must therefore be settled empirically, with consideration as to the many possible dimensions of the neoliberal project and whether they can be traced on the ground. Grounding neoliberalism in the public transport sector of Dar es Salaam in Tanzania required exploring whether the promotion and resistance to neoliberalism was the main game in town or rather an analytical distraction that crowded more salient agendas out of the analysis.

Crucially, making sense of how the public transport system in Dar es Salaam functions, or does not function, and the way in which it has changed over thirty years simply cannot transcend an analytical focus on neoliberalism. Promotion of and resistance to neoliberalism, in complex and different ways over time, can be empirically shown to be the main force behind its development. As we have seen, in the 1980s, 'roll-back' neoliberalism in Dar es Salaam's public transport took the form of the retreat of the public sector from service provision, the privatization of this service, and its progressive economic deregulation. In line with, and as a consequence of, the policy direction of structural adjustment programmes, the public sector had no funds to substantially intervene. With no strong public steer in place, the transport system increasingly became congested and chaotic.

In more recent times, the direction of policy on public transport has conformed to 'roll-out' neoliberalism, manifesting itself in the end of fiscal austerity and in a new focus on boosting the role of the state to reorganize the provision of this service under a public–private partnership, called DART. To this end, a substantial and resourceful unit has been established, with funding from the World Bank. Why should DART be understood as part of a neoliberal project? Answering this question requires examination of the reasons for which the capacity of the public sector was so dramatically boosted. The Tanzanian state has taken on a multi-million dollar debt to overhaul the provision of public transport in Dar es Salaam in a way in which both the infrastructural work, and the service itself, as part of the conditions attached to the World Bank's lending, will not be carried out by the public sector itself. Thus, such an injection of funding is, above all, an initiative to open up public transport in Dar es Salaam to private capital of various forms.

Furthermore, neoliberalism as an ideology, and as a way of seeing the world, is key to understanding the significance of this case study. The influence of neoliberal champion de Soto (1989, 2001) can be seen in the marginalization of the types of workers who are the main protagonists in this case study—workers whose only asset is their own labour—from debates on the informal economy in Africa. This book challenges the idea that micro-scale self-employment is the norm in the informal economy and points to the irrelevance of policy interventions focused on microfinance and the formalization of property rights to the lives of the precarious wage workers on whom I have focused. Instead, for workers of this type interventions should be focused on boosting the demand for labour and on tackling the unregulated nature of the employment relationship in the sector, once more in close alignment with policies associated with neoliberalism.

On the streets of Dar es Salaam, I could indeed identify neoliberal practices, policies, and narratives at work. How and whether they succeeded to influence change was determined by struggles and actual resistance to them, the outcome of which explained the specific form of 'actually existing neoliberalism' in the city transport system. For this reason, the domestic politics and path dependency of the promotion of neoliberalism have taken centre stage in this book: the Tanzanian state attempted, with limited means, to reclaim some policy space on public transport; bus owners resisted marginalization from DART by drawing on pre-existing—and well founded—anxieties in Tanzania about the displacement of national investors by foreign investors; and transport workers have attempted to resist exploitation at work by challenging the unregulated nature of their employment through political organization.

In sum, this case study reminds us that the imagination of alternatives for a better future in African cities must begin with a contextual understanding of the circumstances faced by the urban poor. Such an agenda cannot bypass the study of urban capitalism. It must begin by asking who owns what, who does what with it, who gets power from what, and how is such power deployed to influence change and by grounding neoliberalism and the tensions that it generates in individual places and times.

Questionnaire and Summary of Results

The following questionnaire was prepared and administered by the author in 1998 to a sample of 668 workers.

		Average
1.	*Umri* [Age]	30.2
2.	a) *Dereva* [Driver]	48%
	b) *Kondakta* [Conductor]	52%
3.	*Tangu lini umeanza kufanya shuguliza daladala?* [Since when have you been working with private buses?]	
4.	*Umeshafanya kazi kwenye daladala ngapi?* [How many *daladala* have you worked on?]	1.56 *daladala*/year[1]
5.	*Umeajiriwa kama* [Are you employed]:	
	a) *mkataba?* [with contract?]	17.1%
	b) *kibarua?* [without contract?]	82.9%
6.	*Unafanya kazi kwenye?* [Which make of bus do you work with?]:[2]	
	a) Toyota Hiace	43.9%
	b) Isuzu Coaster	17.8%
	c) Toyota DCM	38.3%
7.	*Kwa mwezi unalipwa kiasi gani?* [What is your monthly wage?]	34,672 shillings

Kwa siku posho kiasi gani unapata? [What is your daily income?]

67.6% answers 'it depends'

32.4% answers, average 2,654 shillings

8. *Kwa wiki unafanya kazi siku ngapi?* [How many days per week do you work?] 6.67 d/w
9. *Kwa siku masaa mangapi ya kazi?* [How many hours per day?]   14.9 h/d
10. *Uhusiano gani na mwenye gari?* [What is your relationship with the owner of the bus?]

a) *Jamaa/Ya kwako* [Relative/own bus]	7.8%
b) *Rafiki* [Friend]	1.7%
c) *Mwenye gari tu* [Purely business]	90.7%

---

[1] This figure is the ratio between the mean of answers to question 3 and that of question 4.

[2] This question aimed to estimate the total number of *daladala employees*. This was possible as both buses (b) and (c) employed three workers each (one driver and two conductors), while (a)—a small bus (for up to a maximum of twenty people)—employed only two workers.

Labour Mobility, December 2001–June 2002

In December 2001 there were twenty-seven buses operating on the route.[1] Over the following six months, seven new buses began operating on the same route, thus bringing fourteen new members to the association. During the same period, eight of the twenty-seven initial buses were not operating on the same route any longer. These buses had either been sold, had broken down, been involved in an accident, or had been moved elsewhere. In the process, sixteen workers had lost their jobs. Of those sixteen, eleven had joined the group of people on the bench, and five had moved elsewhere. Out of the fifty-six people who were on the bench in December 2001, fifteen people were no longer present at the station six months later. Thus, over six months about 15 per cent of the 2001 member base was no longer connected to the association. About 19 per cent were either new entrants (11 per cent), or had changed status, moving from overemployment as *daladala* 'workers with a livelihood' to underemployment on the bench (8 per cent).

Dec. 2001 list of operating buses [27]
    June 2002: 8 of 27 buses OUT
        16 workers 'redundant'
           11 of them became workers on the bench + 5 moved elsewhere
    June 2002: 7 new buses IN
        14 new members
BALANCE OF MEMBERS [December 2001–June 2002]:
    IN: 14 + 11 [on the bench]   = 25
    OUT: 5 + 15 [from the previous list of workers on the bench] = 20
                                        New members + 5
LABOUR FLUIDITY [as percentages] over the 6 months:
    – 19% either new members or in a new category;
    – 15% no longer active members;
    – 34% of people involved in the association have experienced significant change, either as new entrants, being downgraded to the bench, or exiting the association.

---

[1] This appendix presents in fuller detail the data used to justify the step of the argument presented in section 4.6, entitled 'Transport Workers' Horizontal Mobility and its Implications'. See n. 23 in Chapter 4.

Glossary

ajira	employment
bajaj	three-wheeler auto-rickshaw
bodaboda	two-wheeler moto-taxi
daladala	private buses that provide public transport in Dar es Salaam
daladalaman maisha	*daladala* worker 'for life' or also 'with a livelihood'. Slang expression to denote overemployed workers, who spend their life on the bus, as required by their job.
day waka or waka	underemployed workers, who spend most of the day on 'the bench' at *daladala* stations, waiting for workers 'with a livelihood' (see *daladalaman maisha*) to subcontract them part of the working day while having lunch and/or a rest
dereva	driver
hesabu	daily rent paid by *daladala* drivers or conductors to the *daladala* owners
kibarua	work without contract
kijiwe	literally pavement, and in *daladalaman* slang the place where unemployed and underemployed (see *day waka*) workers converge to look for work
kondakta or konda	*daladala* conductor
kukomaa	overripen, agricultural metaphor used by workers to define the rare practice by workers 'with a livelihood' of working all day without subcontracting part of the working day to workers 'on the bench'
kutafuta maisha	to try to build your life, look for a livelihood/job
maisha	life/livelihood
mkataba	contract
mkuu wa reli	literally the head of the railway line, a leader of *daladala* workers, overseeing the organization and running of a *daladala* workers' informal association referred to as the railway line
mshahara	wage

Glossary

ujamaa	Tanzania's socialist experiment
vipanya	small mice, slang for the smallest (eighteen-seater) minibuses
wapiga debe ('those who hit the tin'),	workers who fill the *daladala* buses with passengers at the stops, and who are paid the equivalent of one fare in cash for each bus filled
watu wa benchi	people on the bench (see *day waka*)

References

Adam, C., Cavendish, W., and Mistry, P. S. 1992. *Adjusting Privatisation: Case Studies from Developing Countries*. London: James Currey.

Adarkwa, K. K. and Tamakloe, E. K. A. 2001. 'Urban Transport Problem and Policy Reforms in Kumasi', in K. K. Adarkwa and J. Post (eds), *The Fate of the Tree: Planning and Managing the Development of Kumasi*. Accra, Ghana: Woeli Publishing Services, 139–74.

Adoléhoumé, A. and Bonnafis, A. 2000. 'Urban Transport Micro-Enterprises in Abidjan'. Sub-Saharan Africa Transport Policy Program (SSATP) Technical Note 27. Washington, DC: World Bank.

Adu-Amankwah, K. 1999. 'Ghana', in *Trade Unions in the Informal Sector: Finding their Bearings. Nine Country Papers*. Labour Education Report 116. Geneva: ILO Bureau for Workers' Activities, 1–14.

Agarwala, R. and Herring, R. J. 2008. *Whatever Happened to Class? Reflections from South Asia*. London: Routledge.

Ahferom, M. T. 2009. 'Sustainability Assessment of a Bus Rapid Transit (BRT) System: The Case of Dar es Salaam, Tanzania'. MSc thesis. Lund University, Lund, Sweden.

Alexander, P., Ceruti, C., Motseke, K., Phadi, M., and Wale, K. 2013. 'Class in Soweto', *Africa Today* 60 (1): 134–5.

Aminzade, R. 2013. *Race, Nation and Citizenship in Postcolonial Africa: The Case of Tanzania*. Cambridge: Cambridge University Press.

Amsden, A. H. 2010. 'Say's Law, Poverty Persistence, and Employment Neglect', *Journal of Human Development and Capabilities* 11 (1): 57–66.

Anbalagan, C. and Kanagaraj, K. 2014. 'A Study on Problems and Prospects of Transport in Ethiopia: Special Reference with Auto Rickshaw's [sic] (Bajaj) in Hawassa City, SNNPRS, East Africa', *IFSMRC AIJRM* 2 (3): 26 pp.

Andrae, G. and Beckmann, B. 2010. 'Alliances across the Formal–Informal Divide: South African Debates and Nigerian Experiences', in I. Lindell (ed.), *Africa's Informal Workers: Collective Agency, Alliances and Transnational Organizing in Urban Africa*. London: Zed Books, 85–98.

Ariyo, A. and Jerome, A. 1999. 'Privatization in Africa: An Appraisal', *World Development* 27 (1): 201–13.

Askew, K. M. 2006. 'Sung and Unsung: Musical Reflections on Tanzanian Postsocialisms', Special Issue of *Africa* 76 (1): 15–43.

Azarya, V. and Chazan, N. 1987. 'Disengagement from the State in Africa: Reflections on the Experience of Ghana and Guinea', *Comparative Politics* 29 (1): 106–31.

Bank, L. 1990. 'The Making of the Qwaqwa "Mafia"? Patronage and Protection in the Migrant Taxi Business', *African Studies* 49 (1): 71–93.

Bardasi, E., Beegle, K., Dillon, A., and Serneels, P. 2010. 'Do Labour Statistics Depend on How and to Whom the Questions Are Asked? Results from a Survey Experiment in Tanzania'. Policy Research Working Paper No. 5192. Washington, DC: World Bank.

Baregu, M. 1993. 'The Economic Origins of Political Liberalisation and Future Prospects', in M. S. D. Bagachwa and A. V. Y. Mbelle (eds), *Economic Policy under a Multiparty System in Tanzania*. Dar es Salaam, Tanzania: Dar es Salaam University Press, 105–23.

Barrett, J. 2003. 'Organizing in the Informal Economy: A Case Study of the Minibus Taxi Industry in South Africa'. SEED Working Paper 39, InFocus Programme on Boosting Employment through Small Enterprise Development. Geneva: ILO (International Labour Office).

Bayliss, K. and Fine, B. 2008. 'Privatization in Practice', in Kate Bayliss and Ben Fine (eds), *Privatization and Alternative Public Sector Reform in Sub-Saharan Africa*. London: Palgrave Macmillan, 31–54.

Becker, F. 2013. 'Remembering Nyerere: Political Rhetoric and Dissent in Contemporary Tanzania', *African Affairs* 112 (447): 238–61.

Bennell, P. 1997. 'Privatization in Sub-Saharan Africa: Progress and Prospects during the 1990s', *World Development* 25 (11): 1785–803.

Berg, E. 1999. 'Privatization in Sub-Saharan Africa: Results, Prospects and New Approaches', in J. A. Paulson (ed.), *African Economies in Transition*: vol. 1: *The Changing Role of the State*. London: Palgrave Macmillan, 246–56.

Bernstein, H. 2007. 'Capital and Labour from Centre to Margins'. Prepared for the Living on the Margins Conference, 26–28 March, Stellenbosch, South Africa.

Bernstein, H. 2010. *Class Dynamics of Agrarian Change*. Toronto, Canada: Fernwood Publishing.

Beuving, J. J. 2004. 'Cotonou's Klondike: African Traders and Second-Hand Car Markets in Benin', *Journal of Modern African Studies* 42 (4): 511–37.

Beuving, J. J. 2006. 'Nigerian Second-Hand Car Traders in Cotonou: A Sociocultural Analysis of Economic Decision-Making', *African Affairs* 105 (420): 353–73.

Boampong, O. 2010. 'The Possibilities for Collective Organization of Informal Port Workers in Tema, Ghana', in I. Lindell (ed.), *Africa's Informal Workers: Collective Agency, Alliances and Transnational Organizing in Urban Africa*. London: Zed Books, 130–49.

Bonner, C. and Spooner, D. 2011. 'Organizing in the Informal Economy: A Challenge for Trade Unions', *International Politics and Society* 2: 87–105.

Breman, J. 1996. *Footloose Labour: Working in India's Informal Economy*. Cambridge: Cambridge University Press.

Breman, J. 2003. *The Labouring Poor in India*. New Delhi, India: Oxford University Press.

Breman, J. 2010. *Outcast Labour in Asia: Circulation and Informalization of the Workforce at the Bottom of the Economy*. Oxford: Oxford University Press.

Breman, J. 2013a. 'A Bogus Concept', *New Left Review* 84: 130–8.

Breman, J. 2013b. *At Work in the Informal Economy of India: A Perspective from the Bottom Up*. India: Oxford University Press.

Breman, J. and van der Linden, M. 2014. 'Informalising the Economy: The Return of the Social Question at the Global Level', *Development and Change* 45 (5): 920–40.

Brennan, J. R. 2014. 'Julius Rex: Nyerere through the Eyes of his Critics, 1953–2013', *Journal of Eastern African Studies* 8 (3): 459–77.

Brenner, N., Peck, J., and Theodore, N. 2010. 'Variegated Neoliberalization: Geographies, Modalities, Pathways', *Global Networks* 10 (2): 182–222.

Brenner, N. and Theodore, N. 2002. 'Cities and the Geographies of Actually Existing Neoliberalism', *Antipode* 34 (3): 349–79.

Briggs, J. and Mwamfupe, D. 2000. 'Peri-Urban Development in an Era of Structural Adjustment in Africa', *Urban Studies* 37 (4): 797–809.

Brooks, A. 2012. 'Networks of Power and Corruption: The Trade of Japanese Used Cars to Mozambique', *The Geographical Journal* 178 (1): 80–92.

Brown, A. and Lyons, M. 2010. 'Seen But Not Heard: Urban Voice and Citizenship for Street Traders', in I. Lindell (ed.), *Africa's Informal Workers: Collective Agency, Alliances and Transnational Organizing in Urban Africa*. London: Zed Books, 33–45.

Burawoy, M. and Verdery, K. 1999. 'Introduction', in M. Burawoy and K. Verdery (eds), *Uncertain Transition: Ethnographies of Chance in the Postsocialist World*. Lanham, MD: Rowman and Littlefield, 1–17.

Burton, A. 2005. *African Underclass: Urbanisation, Crime and Colonial Order in Dar es Salaam*. Oxford: James Currey.

Campanelli, P. C., Rothgeb, J. M., and Martin, E. A. 1989. 'The Role of Respondent Comprehension and Interviewer Knowledge in CPS Labour Force Classification'. American Statistical Association, Proceedings of the Section on Survey Research Methods.

Campbell White, O. and Bhatia, A. 1998. *Privatisation in Africa*. Washington, DC: World Bank.

Castells, M. 1996. *The Rise of the Network Society. The Information Age: Economy, Society and Culture*. Oxford: Blackwell.

Castells, M. and Portes, M. 1989. 'World Underneath: The Origins, Dynamics and Effects of the Informal Economy', in A. Portes, M. Castells, and A. Benton (eds), *The Informal Economy: Studies in Advances and Less Developed Countries*. Baltimore: Johns Hopkins University Press, 11–37.

Cervero, R. and Dai, D. 2014. 'BRT TOD: Leveraging Transit Oriented Development with Bus Rapid Transit Investments', *Transport Policy* 36: 127–38.

Chachage, C. and Cassam, A. (eds). 2010. *Africa's Liberation: The Legacy of Nyerere*. Oxford: Pambazuka Press.

Chepchieng, K. J., Kyalo, D. N., and Mulwa, A. S. 2012. *The Influence of Urban Transport Policy on the Growth of Motorcycle and Tricycles in Kenya*. Nairobi, Kenya: University of Nairobi Digital Repository.

Chidaga, E. N. 2010. 'Valuation of Properties of Affected Persons under Resettlement Action Plan of Dar Rapid Transit Project'. Progress Report. Kinondoni Municipal Council.

City of Johannesburg Metropolitan Municipality. 2011. 'Rea Vaya/BRT—Transforming the Face of Public Transport'. End of Term Report 2006–11. Johannesburg, South Africa: City of Johannesburg Metropolitan Municipality.

References

Clarke, S. 2005. 'The Neoliberal Theory of Society', in A. Saad-Filho and D. Johnston (eds), *Neoliberalism: A Critical Reader*. London: Pluto Press, 50–9.

Clerides, S. and Hadjiyannis, C. 2008. 'Quality Standards for Used Durables: An Indirect Subsidy?', *Journal of International Economics* 75: 262–82.

BBC News. 2012. 'Colombia: TransMilenio Bus Protests Paralyse Bogotá'. *BBC News*, 10 March 2012. Available at <http://www.bbc.co.uk/news/world-latin-america-17320247>, accessed 12 December 2013.

Cooksey, B. 2002. 'The Power and the Vainglory: Anatomy of a Malaysian IPP in Tanzania', in K. S. Jomo (ed.), *Ugly Malaysian? South-South Investment Abused*. Durban, South Africa: Centre for Black Research, 47–76.

Cooksey, B. and de Waal, D. 2007. *Why Did City Water Fail? The Rise and Fall of Private Sector Participation in Dar es Salaam's Water Supply*. London: Water Aid.

Cooper, F. 1983. 'Urban Space, Industrial Time and Wage Labour in Africa', in F. Cooper (ed.), *Struggle for the City: Migrant Labor, Capital and the State in Urban Africa*. London: Sage Publications, 7–50.

Cramer, C. 2000. 'Privatisation and Adjustment in Mozambique: A "Hospital Pass"?', *Journal of Southern African Studies* 27 (1): 79–104.

Cramer, C., Johnston, D., Mueller, B., Oya, C., and Sender, J. 2014. 'How to Do (and How Not to Do) Fieldwork on Fair Trade and Rural Poverty', *Canadian Journal of Development Studies* 35 (1): 170–85.

Cramer, C., Oya, C., and Sender, J. 2008. 'Lifting the Blinkers: A New View of Power, Diversity and Poverty in Mozambican Rural Labour Markets', *Journal of Modern African Studies* 46 (3): 361–92.

CTLA, Central Transport Licensing Authority. 1994–8. *Microbuses Files*. Files accessed at UDA. Dar es Salaam, Tanzania: UDA.

Davis, M. 2006. *Planet of Slums*. London and New York: Verso.

DART. 2010. *Resettlement Action Plan for Construction of Dar Rapid Transit (DART) Infrastructure*. Phase 1B report. Dar es Salaam, Tanzania: Dar Rapid Transit Agency, 24–249.

DART. 2014. *BRT System to Contribute to Make Dar a 'Livable' City*. 1 August. Previously available at <www.dart.go.tz/index.php?id=51>, accessed 1 August 2014.

DCC, Dar es Salaam City Council. 1998. *1998 Microbuses File*. Dar es Salaam, Tanzania: DCC.

DCC, Dar es Salaam City Council. 2006. *Conceptual and Detailed Design of a Long Term Integrated Dar es Salaam BRT System*. Draft Final Report on Background on Public Transportation: Annex vol. 2. Dar es Salaam, Tanzania: Logit Engenharia Consultiva.

De Boeck, F. and Plissart, M. F. 2004. *Kinshasa: Tales of the Invisible City*. Leuven, Belgium: Leuven University Press.

Devenish, A. and Skinner, C. 2004. 'Organising Workers in the Informal Economy: The Experience of the Self-Employed Women's Union, 1994–2004'. WIEGO Research Paper. School of Development Studies, University of KwaZulu-Natal. Available at <http://www.streetnet.org.za/docs/research/2004/en/devenishandskinner.pdf>, accessed 1 February 2017.

Dobler, G. 2008. 'From Scotch Whisky to Chinese Sneakers: International Commodity Flows and New Trade Networks in Oshikango, Namibia', *Africa* 78 (3): 410–32.

Doogan, K. 2009. *New Capitalism? The Transformation of Work*. Cambridge: Polity Press.

Dugard, J. 2001. 'Drive On? Taxi Wars in South Africa', in J. Steinberg (ed.), *Crime Wave: The South African Underworld and its Foes*. Johannesburg, South Africa: Witwatersrand University Press, 129–49.

Farlam, P. 2005. 'Working Together: Assessing Public–Private Partnerships in Africa'. Nepad Policy Focus Report No. 2. Pretoria, South Africa: South African Institute for International Affairs, Royal Netherland Embassy.

Ferguson, J. 2015. *Give a Man a Fish: Reflections on the New Politics of Distribution*. Durham, NC and London: Duke University Press.

Fields, G. S. 2005. 'A Guide to Multisector Labour Market Models. Social Protection'. Discussion Paper 0505. Washington, DC: World Bank.

Fine, B. 2008. 'Privatization's Shaky Theoretical Foundations', in K. Bayliss and B. Fine (eds), *Privatization and Alternative Public Sector Reform in Sub-Saharan Africa*. London: Palgrave Macmillan, 13–30.

Fischer, G. 2013. 'Revisiting Abandoned Ground: Tanzanian Trade Unions: Engagement with Informal Workers', *Labour Studies Journal* 38 (2): 139–60.

Fisher, G. 2011. 'Power Repertoires and the Transformation of Tanzanian Trade Unions', *Global Labour Journal* 2 (2): 125–47.

Floro, M. S. and Komatsu, H. 2011. 'Gender and Work in South Africa: What Can Time-Use Data Reveal?', *Feminist Economics* 17 (4): 33–66.

Fox, L. and Pimhidzai, O. 2013. 'Different Dreams, Same Bed: Collecting, Using, and Interpreting Employment Statistics in Sub-Saharan Africa—The Case of Uganda'. Policy Research Working Paper 6436. Washington, DC: World Bank.

Freund, B. 2007. *The African City*. Cambridge: Cambridge University Press.

Gaffey, J. 2015. 'The Challenges Ahead for John Magufuli, Tanzania's Street-Sweeping Leader'. *Newsweek* [Europe], 12 December 2015. Available at <http://europe.newsweek.com/tanzania-challenges-president-john-magufuli-404105>, accessed 12 January 2016.

Gallin, D. 2001. 'Propositions on Trade Unions and Informal Employment in Time of Globalization', *Antipode* 19 (4): 531–49.

Gandy, M. 2005. 'Learning from Lagos', *New Left Review* 33: 37–53.

Ghanadan, R. 2009. 'Connected Geographies and Struggles over Access: Electricity Commercialization in Tanzania', in D. A. McDonald (ed.), *Electricity Capitalism: Recolonising Africa on the Power Grid*. Cape Town, South Africa: HSRC Press and London: Earthscan, 400–36.

Gibbon, P. 1995. 'Merchantisation of Production and Privatisation of Development in Post-*Ujamaa* Tanzania: An Introduction', in P. Gibbon (ed.), *Liberalised Development in Tanzania*. Uppsala, Sweden: Nordiska Afrikainstitutet, 9–36.

Gilbert, A. 2002. 'On the Mystery of Capital and the Myths of Hernando de Soto: What Difference Does Legal Title Make?', *International Development Planning Review* 24 (2): 1–19.

Gilbert, A. 2008. 'Bus Rapid Transit: Is Transmilenio a Miracle Cure?', *Transport Reviews* 28 (4): 439–67.

Godard, X. and Turnier, P. 1992. *Les Transports Urbains en Afrique à l'Heure de l'Ajustement: Redéfinir le Service Public* [*Urban Transport in Africa at the Time of Adjustment: Redefining Public Service*]. Paris: Karthala.

References

Goodfellow, T. and Titeca, K. 2012. 'Presidential Intervention and the Changing "Politics of Survival" in Kampala's Informal Economy', *Cities* 29: 264–70.

Grabher, G. and Stark, D. (eds). 1997. *Restructuring Networks in Post-Socialism: Legacies, Linkages, and Localities*. Oxford: Oxford University Press.

Gray, H. 2012. 'Tanzania and Vietnam: A Comparative Political Economy of Political Transition'. Unpublished PhD dissertation. London: SOAS, University of London.

Gray, H. and Khan, M. 2010. 'Good Governance and Growth in Africa: What Can We Learn from Tanzania?', in V. Padayachee (ed.), *The Political Economy of Africa*. London and New York: Routledge, 339–56.

Guyer, J. I. and Peters, P. 1987. 'Introduction in Conceptualizing the Household: Issues of Theory and Policy in Africa', *Development and Change* 18 (2): 197–214.

Hall, D. 2014. *PPPs*. London: Public Services International Research Unit.

Hann, C. (ed.). 2002. *Postsocialism: Ideals, Ideologies and Practices in Eurasia*. London: Routledge.

Hansen, K. T. and Vaa, M. (eds). 2004. *Reconsidering Informality: Perspectives from Urban Africa*. Uppsala, Sweden: Nordiska Afrikainstitutet.

Harrison, G. 2010. *Neoliberalism in Africa: The Impact of Global Social Engineering*. London and New York: Zed Books.

Harriss-White B. 2003. *India Working: Essays on Society and Economy*. Cambridge: Cambridge University Press.

Harriss-White B. 2010. 'Work and Wellbeing in Informal Economies: The Regulative Roles of Institutions of Identity and the State', *World Development* 38 (2): 170–83.

Harriss-White, B. and Gooptu, N. 2000. 'Mapping India's World of Unorganized Labour', in Leo Panitch and Colin Leys (eds), *Socialist Register 2001: Working Classes, Global Realities*. London: Merlin Press, 89–118.

Hart, G. 2008. 'The Provocations of Neoliberalism: Contesting the Nation and Liberation after Apartheid', *Antipode* 40 (4): 678–705.

Hart, K. 1973. 'Informal Income Opportunities and Urban Employment in Ghana', *Journal of Modern African Studies* 11 (1): 61–89.

Harvey, D. 2005. *A Brief History of Neoliberalism*. Oxford: Oxford University Press.

Hidalgo, D., Custodio, P., and Graftieaux, P. 2007. 'A Critical Look at Major Bus Improvements in Latin America and Asia: Case Studies of Hitches, Hic-Ups [sic] and Areas for Improvement: Synthesis of Lessons Learned'. Washington, DC: World Bank.

Hidalgo, D. and Gutiérrez, L. 2013. 'BRT and BLHS around the World: Explosive Growth, Large Positive Impacts and Many Issues Outstanding', *Research in Transportation Economics* 39: 8–13.

Home, R. and Lim, H. 2004. *Demystifying the Mystery of Capital: Land Tenure and Poverty in Africa and the Caribbean*. London: Glass House Press.

IBIS Transport Consultants Ltd. 2005. 'Study of Urban Public Transport Conditions in Accra, Ghana'. Accra, Ghana: Public Private Infrastructure Advisory Facility.

ICFTU, International Confederation of Free Trade Unions. 2006. 'Core Labour Standards in Tanzania'. Report for the WTO General Council Review of the Trade Policies of Tanzania. Geneva, Switzerland. Previously available at <http://www.icftu.org/www/pdf/corelabourstandards2006tanzania.pdf>, accessed 23 March 2012.

IDL. 2013. 'Disordered Cities: Analysing Governance in the Urban Transport Sector'. IDL Working Paper No. 2. Dar es Salaam, Tanzania, and Dakar, Senegal: IDL.

IEA, International Energy Agency. 2002. 'Bus Systems for the Future: Achieving Sustainable Transport Worldwide'. Paris: OECD.

Iliffe, J. 1979. *A Modern History of Tanganyika*. Cambridge: Cambridge University Press.

ILO, International Labour Organization. 2013. 'Measuring Informality: A Statistical Manual on the Informal Sector and Informal Employment'. Geneva: ILO.

ITDP, Institute for Transportation and Development Policy. 2011. *Annual Report 2011*. Available at <https://www.itdp.org/wp-content/uploads/2014/07/ITDP_AR_11_9. 13.12.pdf>, accessed 1 February 2017.

ITF, International Transport Workers Federation. 2010. 'Transport Toolkit: Trade Union Influence on World Bank Projects. Case Study: Urban Public Transport in Bogota'. London: ITF.

Jamal, V. and Weeks, J. 1993. *Africa Misunderstood: Or Whatever Happened to the Rural–Urban Gap?* Basingstoke: Macmillan.

Jason, A. 2008. 'Organizing Informal Workers in the Urban Economy: The Case of the Construction Industry in Dar es Salaam, Tanzania', *Habitat International* 32 (2): 192–202.

Jerven, M. 2013. *Poor Numbers: How We Are Misled by African Development Statistics and What to Do about It*. Ithaca, NY and London: Cornell University Press.

Jerven, M. and Johnston, D. 2015. 'Statistical Tragedy in Africa? Evaluating the Data Base for African Economic Development', Special Issue of *The Journal of Development Studies* 51 (2): 111–15.

Jessop, B. 2002. 'Liberalism, Neoliberalism and Urban Governance: A State-Theoretical Perspective', *Antipode* 34 (3): 452–72.

Jessop, B. 2013. 'Putting Neoliberalism in its Time and Place: A Response to the Debate', *Social Anthropology* 21 (1): 65–74.

JICA, Japan International Cooperation Agency. 1995. 'The Study of Dar es Salaam Road Development Plan'. Final Report.

Jimu, I. M. 2010. 'Self-Organized Informal Workers and Trade Union Initiatives in Malawi: Organizing in the Informal Economy', in I. Lindell (ed.), *Africa's Informal Workers: Collective Agency, Alliances and Transnational Organizing in Urban Africa*. London: Zed Books, 99–114.

Jonsson, J. B. and Bryceson, D. F. 2009. 'Rushing for Gold: Mobility and Small-Scale Mining in East Africa', *Development and Change* 40 (2): 249–79.

Jonsson, J. B. and Fold, N. 2009. 'Handling Uncertainty: Policy and Organizational Practices in Tanzania's Small-Scale Gold Mining Sector', *Natural Resources Forum* 33: 211–20.

Ka'bange, A., Mfinanga, D., and Hema, E. 2014. 'Paradoxes of Establishing Mass Rapid Transit Systems in African Cities: A Case of Dar es Salaam Rapid Transit (DART) System, Tanzania', *Research in Transportation Economics* 48: 76–183.

Kabeer, N., Sudarshan, R. M., and Milward, K. 2013. *Organizing Women Workers in the Informal Economy: Beyond the Weapons of the Weak*. London and New York: Zed Books.

Kamukala, G. L. 2010. 'Daladala Grievances Plan'. Unpublished consultancy, prepared for DART, and in author's possession. Dar es Salaam, Tanzania.

References

Kanyama, A., Carlsson-Kanyama, A., Linden A. L., and Lupala, J. 2004. 'Public Transport in Dar es Salaam, Tanzania: Institutional Challenges and Opportunities for a Sustainable Transportation System'. Stockholm: Royal Institute of Technology.

Kassim, A. and Ali, M. 2006. 'Solid Waste Collection by the Private Sector: Households' Perspective—Findings from a Study in Dar es Salaam City, Tanzania', *Habitat International* 30: 769–80.

Kelsall, T. 2003. 'Governance, Democracy and Recent Political Struggles in Mainland Tanzania', *Commonwealth and Comparative Politics* 41 (2): 55–82.

Kelsall, T. 2013. *Business, Politics and the State in Africa: Challenging the Orthodoxies on Growth and Transformation*. London and New York: Zed Books.

Khayesi, M. 1998. 'The Need for an Integrated Road Safety Programme for the City of Nairobi, Kenya', in CODATU Association, P. Freeman, and C. Jamet (eds), *Urban Transport Policy: A Sustainable Development Tool: Proceedings of the International Conference CODATU VIII, Cape Town, South Africa, 21–25 September 1998*. Rotterdam: A. A. Balkema, 579–82.

Khosa, M. M. 1992. 'Routes, Ranks and Rebels: Feuding in the Taxi Revolution', *Journal of Southern African Studies* 18 (1): 232–51.

Khosa, M. M. 1994. 'Accumulation and Labour Relations in the Taxi Industry', *Transformations* 24: 55–71.

Kironde, L. J. M. 1999. 'The Governance of Waste Management in African Cities', in A. Atkinson et al. (ed.), *The Challenge of Environmental Management in Urban Areas*. Aldershot: Ashgate, 75–88.

Kisyombe, D. N. 2005. 'Drug Use among Commuter Bus Operators: A Case Study of Dar es Salaam City'. Dissertation for Postgraduate Diploma in Social Work. University of Dar es Salaam, Dar es Salaam, Tanzania.

Kjær, A. M. and Therkildsen, O. 2013. 'Elections and Landmark Policies in Tanzania and Uganda', *Democratization* 20 (4): 592–614.

Kombe, W., Kyessi, A., Lupala, J., and Mgonia, E. 2003. 'Partnership to Improve Access and Quality of Public Transport. A Case Report: Dar es Salaam, Tanzania'. Leicestershire: Water, Engineering and Development Centre, Loughborough University. Available at <https://dspace.lboro.ac.uk/dspace-jspui/bitstream/2134/9566/2/Partnerships_to_improve_access_transport_-_Tanzania.pdf>, accessed 1 February 2017.

Kouakou, E. and Fanny, D. 2008. 'Overview of Public Transport in Sub-Saharan Africa'. Trans-Africa Project of European Commission. Joint Report of UITP—International Association of Public Transport, and UATP—African Association of Public Transport. Brussels, Belgium: UITP, International Association of Public Transport.

Krueckeberg, D. 2004. 'The Lessons of John Locke or Hernando de Soto: What if Your Dreams Come True?', *Housing Policy Debate* 15 (1): 1–24.

Kulaba, S. 1989. 'Local Government and the Management of Urban Services', in R. Stren and R. R. White (eds), *African Cities in Crisis: Managing Rapid Urban Growth*. Boulder, CO: Westview Press, 203–46.

Kumar, A. and Barrett, F. 2008. *Stuck in Traffic: Urban Transport in Africa*. AICD—Africa Infrastructure Country Diagnostic Project, implemented by the World Bank, January 2008.

Kyessi, A. G. 2005. 'Community-Based Urban Water Management in Fringe Neighbourhoods: The Case of Dar es Salaam, Tanzania', *Habitat International* 29 (1): 1–25.

Lambright, G. M. S. 2014. 'Opposition Politics and Urban Service Delivery in Kampala', *Development Policy Review* 32 (1): 39–60.

Lal, P. 2012. 'Self-Reliance and the State: The Multiple Meanings of Development in Early Post-Colonial Tanzania', *Africa: Journal of the International African Institute* 82 (2): 212–34.

Lange, S. 2011. 'Gold and Governance: Legal Injustices and Lost Opportunities in Tanzania', *African Affairs* 110 (439): 233–52.

LeBrun, O. and Gerry, C. 1974. 'Petty Producers and Capitalism', *Review of African Political Economy* 1 (2): 20–32.

Lerche, J. 2012. 'Labour Regulations and Labour Standards in India: Decent Work?', *Global Labour Journal* 3 (1): 16–39.

Li, T. 2010. 'To Make Live or Let Die? Rural Dispossession and the Protection of the Surplus Populations', *Antipode* 41 (supplement S1): 66–93.

Lindell, I. (ed.). 2010. *Africa's Informal Workers: Collective Agency, Alliances and Transnational Organizing in Urban Africa*. London: Zed Books.

Locatelli, F. and Nugent, P. 2009. 'Introduction', in F. Locatelli and P. Nugent (eds), *African Cities: Competing Claims on Urban Spaces*. Leiden, Netherlands: Kroninklijke Brill NV, 1–13.

Lourenco-Lindell, I. 2002. *Walking the Tight Rope: Informal Livelihoods and Social Networks in a West African City*. Stockholm Acta Universitatis Stockholmiensis, Stockholm Studies in Human Geography 9. Stockholm: Almqvist & Wiksell International.

Loxley, J. 2013. 'Are Public–Private Partnerships (PPPs) the Answer to Africa's Infrastructure Needs?', *Review of African Political Economy* 40 (137): 485–95.

Lugalla, J. P. L. 1995. *Crisis, Urbanization and Urban Poverty in Tanzania: A Study of Urban Poverty and Survival Politics*. Lanham, MD: University Press of America.

MacGaffey, J. 1991. *The Real Economy of Zaire: The Contribution of Smuggling and Other Unofficial Activities to National Wealth*. London: James Currey and Philadelphia, PA: University of Pennsylvania Press.

Maliyamkono, T. L. and Bagachwa, M. S. D. 1990. *The Second Economy in Tanzania*. London: James Currey.

Mamdani, M. 1996. *Citizen and Subject: Contemporary Africa and the Legacy of Colonialism*. Princeton, NJ: Princeton University Press.

Mamuya, I. 1993. *Structural Adjustment and the Reform of the Public Sector Control System in Tanzania*. Hamburg: Institut für Afrika-Kunde.

Marcel, B. and Ngewe, G. 2011. 'Tumbling the Impenetrable Road Congestion in Dar es Salaam: A Commuter Service Perspective', in *Proceedings of 26th National Conference on Challenges in Addressing Traffic Congestion and Enhancing Road Safety for National Development, December 1–2, Arusha, Tanzania*. Tanzania: Institution of Engineers, 31–9.

Martin, E. and Polivka, A. E. 1995. 'Diagnostics for Redesigning Survey Questionnaires', *Public Opinion Quarterly* 59: 547–67.

Marx, K. 1852/1974. 'The Eighteenth Brumaire of Louis Bonaparte', in D. Fernbach (ed.), *Surveys from Exile: Political Writings*, vol. II. New York: Vintage, 146–249.

References

Marx, K. 1867/1976. *Capital*, vol 1. Harmondsworth, UK: Penguin.

Marx, K. and Engels, F. 1844/2012. *The Holy Family*. Honolulu, HI: University Press of the Pacific.

Matsumoto, N. 2006. 'Analysis of Policy Processes to Introduce Bus Rapid Transit Systems in Asian Cities from the Perspective of Lesson-Drawing: Cases of Jakarta, Seoul, and Beijing'. Paper presented at the Better Air Quality Workshop. Yogyakarta, Indonesia.

McDonald, D. A. and Sahle, E. N. (eds). 2002. *The Legacies of Julius Nyerere: Influences on Development Discourse and Practice in Africa*. Trenton, NJ: Africa World Press.

McKinsey Global Institute. 2010. 'Lions on the Move: The Progress and Potential of African Economies'. Report, June 2010. McKinsey Global Institute.

MCT, Ministry of Communication and Transport. 1987. *Proposed National Transport Policy*. Dar es Salaam, Tanzania: MCT.

MCT, Ministry of Communication and Transport. 1998. 'Press Release of the Official Meeting about Students' Harassment by *Daladala* Operators'. Dar es Salaam, Tanzania: MCT.

Meagher, K. 2010a. *Identity Economics: Social Networks and the Informal Economy in Nigeria*. Woodbridge, Suffolk and Rochester, NY: James Currey and Ibadan, Nigeria: HEBN Publishers.

Meagher, K. 2010b. 'The Politics of Vulnerability: Exit Voice and Capture in Three Nigerian Informal Manufacturing Clusters', in I. Lindell (ed.), *Africa's Informal Workers: Collective Agency, Alliances and Transnational Organizing in Urban Africa*. London: Zed Books, 46–64.

Mfinanga, D. and Madinda, E. 2016. 'Public Transport and Daladala Service Improvement Prospects in Dar es Salaam', in R. Beherens, D. McCormick, and D. Mfinanga (eds), *Paratransit in African Cities: Operations, Regulation and Reform*. Abingdon, UK: Routledge, 155–73.

Mitulla, W. V. and Wachira, I. N. 2003. 'Informal Labour in the Construction Industry in Kenya: A Case Study of Nairobi'. ILO Sectoral Activities Programme, Working Paper 204. Geneva: International Labour Organization.

Mkandawire, T. 2001. 'Thinking about Developmental States in Africa', *Cambridge Journal of Economics* 25: 289–313.

Mkandawire, T. and Soludo, C. C. 1998. *Our Continent, our Future: African Perspectives on Structural Adjustment*. Dakar, Senegal: CODESRIA, Ottawa, Canada: IDRC Books, and Asmara, Eritrea: Africa World Press.

Monson, J. 2006. 'Defending the People's Railway in the Era of Liberalization: TAZARA in Southern Tanzania', *Africa* 76 (1): 113–30.

Mrema, G. D. 2011. 'Traffic Congestion in Tanzania Major Cities: Causes, Impacts and Suggested Mitigations to the Problem', in *Proceedings of 26th National Conference on Challenges in Addressing Traffic Congestion and Enhancing Road Safety for National Development, December 1–2, Arusha, Tanzania*. Tanzania: Institution of Engineers, 1–20.

Mudge, S. 2008. 'What is Neo-Liberalism?', *Socio-Economic Review* 6 (4): 703–31.

Muhajir, M. 2011. 'How Planning Works in an Age of Reform: Land, Sustainability, and Housing Development Traditions in Zanzibar'. PhD Thesis. Kansas African Studies Centre, University of Kansas, Kansas City, KA.

Munck, R. 2013. 'The Precariat: A View from the South', *Third World Quarterly* 34 (5): 747–62.

Muñoz, J. C., Batarce, M., and Hidalgo, D. 2013. *Transantiago, Five Years after its Launch*. Santiago, Chile: Department of Transport Engineering and Logistics, Pontificia Universidad Católica de Chile.

Mutasingwa, D. S. n.d. 'Public Transport Reform in Tanzania: Dar es Salaam City as a Case Study'. Consultancy Report for the Institute for Transportation and Development.

Mutongi, K. 2006. 'Thugs or Entrepreneurs? Perceptions of Matatu Operators in Nairobi, 1970 to the Present', *Africa* 76 (4): 549–68.

Myers, G. 2008. 'Peri-Urban Land Reform, Political Economy Reform, and Urban Political Ecology in Zanzibar', *Urban Geography* 29 (3): 264–88.

Myers, G. 2011. *African Cities: Alternative Visions of Urban Theory and Practice*. London and New York: Zed Books.

Myrdal, G. 1968. *Asian Drama*, vol. 3. London: Penguin.

Narayan, L. and Chikarmane, P. 2013. 'Power at the Bottom of the Heap: Organizing Waste Pickers in Pune', in N. Kabeer et al. (eds), *Organizing Women Workers in the Informal Economy: Beyond the Weapons of the Weak*. London: Zed Books, 205–31.

National Institute for Transport. 2010. 'Report on the Participation of Local Bus and Truck Operators in the DART Project'. Dar es Salaam, Tanzania: DART.

NBS, National Bureau of Statistics. 1989. '1988 Population Census: Preliminary Report'. Dar es Salaam, Tanzania: Ministry of Finance, Planning and Economic Affairs.

NBS, National Bureau of Statistics. 2002. 'Household Budget Survey 2000/01'. Dar es Salaam, Tanzania: National Bureau of Statistics.

NBS, National Bureau of Statistics. 2007. 'Integrated Labour Force Survey (ILFS) 2006'. Analytical Report. Dar es Salaam, Tanzania: National Bureau of Statistics.

NBS, National Bureau of Statistics. 2009a. 'ILFS 2006, Labour Force Survey FORM 1'. Dar es Salaam, Tanzania: National Bureau of Statistics.

NBS, National Bureau of Statistics. 2009b. 'ILFS 2006, Labour Force Survey FORM 2'. Dar es Salaam, Tanzania: National Bureau of Statistics.

NBS, National Bureau of Statistics. 2009c. 'ILFS 2006, Dodoso la CLS 1'. Dar es Salaam, Tanzania: National Bureau of Statistics.

NBS, National Bureau of Statistics, 2013. '2012 Population Census'. Dar es Salaam, Tanzania: National Bureau of Statistics.

NRT, National Institute for Transport. 2010. 'Report on the Participation of Local Bus and Truck Operators in the DART Project'. Dar es Salaam, Tanzania: DART.

Ntambara, A. 2013. 'Unreliable Transport: A Problem for Students in Dar es Salaam'. Available at <https://wewriteforrights.wordpress.com/2013/08/22/unreliable-transport-a-problem-for-students-in-dar-es-salaam/>, accessed 9 April 2015.

Nuttall, S. and Mbembe, A. 2005. 'A Blasé Attitude: A Response to Michael Watts', *Public Culture* 17 (1): 193–202.

O'Connor, A. 1988. 'The Rate of Urbanisation in Tanzania in the 1970s', in M. Hood (ed.), *Tanzania after Nyerere*. London: Pinter, 136–42.

Olukoshi, A. O. 2003. 'The Elusive Prince of Denmark: Structural Adjustment and the Crisis of Governance in Africa', in T. Mkandawire and C. C. Soludo (eds), *African*

Voices on Structural Adjustment. Ottawa, Canada: IDRC Books and Dakar, Senegal: CODESRIA, 220–73.

Oya, C. 2013. 'Rural Wage Employment in Africa: Methodological Issues and Emerging Evidence', *Review of African Political Economy* 40 (136): 251–73.

Paget-Seekins, L. 2015. 'Bus Rapid Transit as a Neoliberal Contradiction', *Journal of Transport Geography* 48: 115–20.

Paget-Seekins, L., Flores, O., and Muñoz, J. C. 2015. 'Examining Regulatory Reform for Bus Operations in Latin America', *Urban Geography* 36 (3): 424–38.

Phelps, N. A., Power, M., and Wanjiru, R. 2007. 'Learning to Compete: Communities of Investment Promotion Practice in the Spread of Global Neoliberalism', in K. England and K. Ward (eds), *Neoliberalization: States, Networks and People*. Oxford and Malden, MA: Blackwell, 370–412.

Pieterse, E. 2008. *City Futures: Confronting the Crisis of Urban Development*. London and New York: Zed Books.

Pirie, G. 2014. 'Transport Pressures in Urban Africa: Practices, Policies, Perspectives', in S. Parnell and E. Pieterse (eds), *Africa's Urban Revolution*. London and New York: Zed Books, 133–47.

Pitcher, A. M. and Askew, K. M. 2006. 'African Socialism and Postsocialism', Special Issue of *Africa* 76 (1): 1–14.

Pitcher, A. M. and Graham, A. 2006. 'Cars Are Killing Luanda: Cronyism, Consumerism, and Other Assaults on Angola's Postwar Capital City', in M. J. Murray and G. Myers (eds), *Cities in Contemporary Africa*. New York: Palgrave Macmillan, 173–94.

Piven, F. F. and Cloward, R. A. 2000. 'Power Repertoires and Globalization', *Politics and Society* 28 (3): 413–30.

Ponte, S. 2004. 'The Politics of Ownership: Tanzanian Coffee Policy in the Age of Liberal Reformism', *African Affairs* 103 (413): 615–33.

Porter, J. 2010. 'Trade Union Responses to World Bank Restructuring Projects: The Case of Transmilenio in Colombia'. Paper prepared for Public World Workshop, London, June 2010. Available at <http://www.publicworld.org/files/colombiabrtenglish.pdf>, accessed 21 January 2014.

Portes, A., Castells, M., and Benton, L. A. 1989. *The Informal Economy: Studies in Advanced and Less Developed Countries*. Baltimore and London: Johns Hopkins University Press.

Quality Public Transport. 2012. 'A Model BRT? Transmilenio in Bogotá'. *Quality Public Transport*, Briefing No. 8: 2 pp.

Quinn, S. and Teal, F. 2008. 'Private Sector Development and Income Dynamics: A Panel Study of the Tanzanian Labour Market'. CSAE Working Paper Series 2008–09. Oxford: Centre for the Study of African Economies.

Radice, H. 2014. 'Class Theory and Class Politics Today', in L. Panitch and G. Albo (eds), *Socialist Register 2015: Transforming Classes*. London: Merlin Press, 270–92.

Randall, S. and Coast, E. 2015. 'Poverty in African Households: The Limits of Survey Representations', *The Journal of Development Studies* 51 (2): 162–77.

Rasmussen, J. 2012. 'Inside the System, Outside the Law: Operating the Matatu Sector in Nairobi', *Urban Forum* 23 (4): 415–32.

Rebel Group. 2014. *DART Project Phase 1: Project Information Memorandum—Final Version*. Report prepared for the DART Agency. Washington, DC: World Bank.

Resnik, D. 2012. 'Opposition Parties and the Urban Poor in African Democracies', *Comparative Political Studies* 45 (11): 1351–78.

Rizzo, M. 2002. 'Being Taken for a Ride: Privatisation of the Dar es Salaam Transport System 1983–1998', *Journal of Modern African Studies* 40 (1): 133–57.

Rizzo, M. 2011. '"Life is War"! Informal Transport Workers and Neoliberalism in Tanzania, 1998–2009', *Development and Change* 42 (5): 179–206.

Rizzo, M. 2013. 'Informalisation and the End of Trade Unionism as We Knew It? Dissenting Remarks from a Tanzanian Case Study', *Review of African Political Economy* 40 (136), 290–308.

Rizzo, M. 2015. 'The Political Economy of an African Urban Megaproject: The Bus Rapid Transit Project in Tanzania', *African Affairs* 114 (455): 249–70.

Rizzo, M., Kilama, B., and Wuyts, M. 2015. 'The Invisibility of Wage Employment in Statistics on the Informal Economy in Africa: Causes and Consequences', *The Journal of Development Studies* 51: 149–61.

Robinson, J. 2006. *Ordinary Cities: Between Modernity and Development*. Questioning Cities Series. Abingdon, UK: Routledge.

Robinson, J. and Parnell, S. 2011. 'Travelling Theory: Embracing Post-Neoliberalism through Southern Cities', in G. Bridge and S. Watson (eds), *The New Blackwell Companion to the City*. Oxford: Blackwell, 521–31.

Saad-Filho, A. and Johnston, D. 2005. 'Introduction', in A. Saad-Filho and D. Johnston (eds), *Neoliberalism: A Critical Reader*. London: Pluto Press, 1–6.

Salon, D. and Gulyani, S. 2010. 'Mobility, Poverty and Gender: Travel "Choices" of Slum Residents', *Transport Reviews: A Transnational Transdisciplinary Journal* 30 (5): 641–57.

Sanders, T. 2008. 'Buses in Bongoland: Seductive Analytics and the Occult', *Anthropological Theory* 8: 107–32.

dos Santos, A. 2005. 'A viaturolatria', *Agora* 426 (May 28): 3.

Sarris, A. H. and Van den Brink, R. 1993. *Economic Policy and Household Welfare during Crisis and Adjustment in Tanzania*. New York and London: New York University Press.

Satterwhite, D. 2006. 'Book Reviews. Planets of Slums and Shadow Cities', *Environment and Urbanization* 18 (2): 543–6.

Schakelamp, H. 2011. 'Reflecting on Progress with Cape Town's MyCiTi Bus System', *Mobility* (January/April): 10–12.

Schakelamp, H. and Beherens, R. 2010. 'Engaging Paratransit on Public Reform Initiatives in South Africa: A Critique of Policy and an Investigation of Appropriate Engagement Approaches', *Research in Transport Economics* 29: 371–8.

Schakelamp, H. and Beherens, R. 2013. 'Engaging the Paratransit Sector in Cape Town on Public Transport Reform: Progress, Process and Risks', *Research in Transport Economics* 39: 185–90.

Selwyn, B. 2007. 'Labour Process and Workers' Bargaining Power in Export Grape Production, North East Brazil', *Journal of Agrarian Change* 7 (4): 526–53.

Sender, J., Oya, C., and Cramer, C. 2006. 'Women Working for Wages: Putting Flesh on the Bones of a Rural Labour Market Survey in Mozambique', *Journal of Southern African Studies* 32 (2): 313–33.

Shivji, I. G. 1986. *Law, State, and the Working Class in Tanzania, c. 1920–1964*. London: James Currey and Portsmouth, NH: Heinemann.

Silver, B. J. 2003. *Forces of Labour: Workers' Movements and Globalization since 1870*. Cambridge: Cambridge University Press.

Simone, A. 2004a. 'People as Infrastructure: Intersecting Fragments in Johannesburg', *Public Culture* 16 (3): 407–29.

Simone, A. 2004b. *For the City Yet to Come: Changing African Lives in Four Cities*. Durham, NC: Duke University Press.

Simone, A. 2005. 'Introduction: Urban Processes and Change', in A. Simone and A. Abouhani (eds), *Urban Africa: Changing Contours of Survival in the City*. Dakar, Senegal: CODESRIA and London: Zed Books, 1–26.

Simone, A. 2010. *City Life from Jakarta to Dakar: Movements at the Crossroads*. London: Routledge.

de Soto, H. 1989. *The Other Path*. New York: Harper and Row.

de Soto, H. 2001. *The Mystery of Capital: Why Capitalism Triumphs in the West and Fails Everywhere Else*. London: Black Swan.

Standing, G. 2006. 'Labour Markets', in D. A. Clark (ed.), *The Elgar Companion to Development Studies*. Cheltenham: Edward Elgar, 323–8.

Standing, G. 2011. *The Precariat: The New Dangerous Class*. London: Bloomsbury Academic.

Stein, H. 1991. 'Economic Policy and the IMF in Tanzania: Conditionality, Conflict and Convergence', in H. Campbell and H. Stein (eds), *The IMF and Tanzania*. Harare, Zimbabwe: Southern Africa Political Economy Series Trust, 86–113.

Stenning A. C. 2005. 'Post-Socialism and the Changing Geographies of the Everyday in Poland', *Transactions of the Institute of British Geographers* 30 (1): 113–27.

Stewart, A. 2014. 'BRT in Dar es Salaam'. Available at <http://ansoncfit.com/watson/brt-in-dar-es-salaam/>, accessed 1 February 2017.

Storper, M. and Scott, A. J. 2016. 'Current Debates in Urban Theory: A Critical Assessment', *Urban Studies* 53 (6): 1114–36.

Stren, R. 1989. 'The Administration of Urban Services', in R. Stren and R. R. White (eds), *African Cities in Crisis: Managing Rapid Urban Growth*. Boulder, CO: Westview Press, 37–68.

SUMATRA, Surface and Marine Transport Regulatory Authority. 2007. 'Study on Student Transport Problems in Dar es Salaam and Other Urban Centres in Mainland Tanzania'. Final Report. Dar es Salaam, Tanzania: National Institute of Transport.

SUMATRA, Surface and Marine Transport Regulatory Authority. 2009. '*Daladala* Information'. Unpublished document in author's possession, 2 pp.

SUMATRA, Surface and Marine Transport Regulatory Authority. 2011. 'Study on User Need and Management of Public Transport Services in Dar es Salaam'. Dar es Salaam, Tanzania: SUMATRA.

Tambulasi, R. I. C. and Kayuni, H. M. 2008. 'Can the State Perpetuate the Marginalisation of the Poor? The Socio-Economic Effects of the State's Ban on Minibus "Callboys" in Malawi', *Development Southern Africa* 25 (2): 215–26.

Tanzania Invest. 2016. 'Tanzania Assigns New CEO to Dar es Salaam Rapid Transit Agency DART', 22 January. Available at <http://www.tanzaniainvest.com/transport/tanzania-assigns-new-ceo-to-dar-es-salaam-rapid-transit-agency-dart>, accessed 16 March 2016.

Tati, G. 2001. 'Responses to the Urban Crisis in Cameroon and Congo: Patterns of Local Participation in Urban Management', in A. Tostensen, I. Tvedten, and M. Vaa (eds), *Associational Life in African Cities: Popular Responses to the Urban Crisis*. Uppsala, Sweden: Nordiska Afrikainstitutet, 182–97.

Temu, A. E. and Due, J. M. 2000. 'The Business Environment in Tanzania after Socialism: Challenges of Reforming Banks, Parastatals, Taxation and the Civil Service', *Journal of Modern African Studies* 38 (4): 683–712.

Tomic, P. and Trumper, R. 2005. 'Powerful Drivers and Meek Passengers: On the Buses in Santiago', *Race and Class* 47 (1): 49–63.

TRA, Tanzania Revenue Authority. 1998a. *Daladala Files* (District of Ilala). Dar es Salaam, Tanzania: TRA.

TRA, Tanzania Revenue Authority. 1998b. *Daladala Files* (District of Kinondoni). Dar es Salaam, Tanzania: TRA.

TRA, Tanzania Revenue Authority. 1998c. *Daladala Files* (District of Temeke). Dar es Salaam, Tanzania: TRA.

Trebilcock, M. and Rosenstock, M. 2015. 'Infrastructure Public–Private Partnerships in the Developing World: Lessons from Recent Experience', *The Journal of Development Studies* 51 (4): 335–54.

Tripp, A. M. 1997. *Changing the Rules: The Politics of Liberalization and the Urban Informal Economy in Tanzania*. London and Berkeley, CA: University of California Press.

Turner, R. 2008. *Neo-Liberal Ideology*. Edinburgh: Edinburgh University Press.

UDA leaflet. n.d. 'For Better Transport'. Available at <http://www.jamiiforums.com/great-thinkers/154281-uda-imebinafsishwa-ili-iingie-biashara-na-dart-siri-ya-kuuzwa-bei-chee.html>, accessed 4 August 2014.

UDA, Shirika la Usafiri Dar es Salaam. 1994. 'Five Year Development Plan'. Unpublished Report. Dar es Salaam, Tanzania: UDA.

UDA, Shirika la Usafiri Dar es Salaam. 1994–8. 'Fleet Performance Files'. Annual files from 1994/5 to 1997/8. Dar es Salaam, Tanzania: UDA.

UDA, Shirika la Usafiri Dar es Salaam. 1995. 'Maelezo mafupi kuhusu UDA na shuguli zake yasiyowasolishwa kwa mheshimiwa Dr Maua Daftari Mbunge Naibu Waziri wa Mawasiliano na Uchukuzi alipotembelea UDA 21.12.1995' (A short report on the performance of UDA, presented by Dr Maua, Vice-Minister of Communication and Transport, on the occasion of his visit to UDA). Unpublished Report. Dar es Salaam, Tanzania: UDA.

UNDP and United Republic of Tanzania. 2015. 'Tanzania Human Development Report 2014: Economic Transformation for Human Development'. Dar es Salaam, Tanzania: Economic and Social Research.

URT, United Republic of Tanzania. 1974. *Exclusive Licence Granted to Shirika la Usafiri Dar es Salaam to Operate Passenger Services in Dar es Salaam*. Unpublished licence, 19 January. Dar es Salaam, Tanzania.

URT, United Republic of Tanzania. 1996. *Road Traffic Act Amendment, no. 16*. Dar es Salaam, Tanzania.

URT, United Republic of Tanzania. 2001. *The Surface and Marine Transport Regulatory Authority Act*. Dar es Salaam, Tanzania: MPP.

URT, United Republic of Tanzania. 2007. *The Transport Licencing (Road Passenger Vehicles) Regulations*. Government Notice No. 218, 26 October. Dar es Salaam, Tanzania: MPP.

URT, United Republic of Tanzania. 2008. Government Notice No. 193, 18 April. Dar es Salaam, Tanzania: Tanzania Government Printer.

URT, United Republic of Tanzania. 2009. 'National Road Safety Policy'. Dar es Salaam, Tanzania: Ministry of Infrastructure Development.

Utz, R. K. (ed.). 2008. *Sustaining and Sharing Economic Growth in Tanzania*. Washington, DC: World Bank.

UWAMADAR. 2003. 'Working Environment of Dar es Salaam Commuting Bus Drivers and Conductors. Final Report'. November 2003. Dar es Salaam, Tanzania: Konrad Adenauer Stiftung and Development Dynamics International.

UWAMADAR. 2010. 'Drinking and Driving among *Daladala* Drivers. Baseline Data Survey: Final Report'. Dar es Salaam, Tanzania: Konrad Adenauer Stiftung and Development Dynamics International.

Van Waeyenberge, E., Fine, B., and Bayliss, K. 2011. 'The World Bank, Neoliberalism and Development Research', in K. Bayliss, E. Van Waeyenberge, and B. Fine (eds), *The Political Economy of Development: The World Bank, Neoliberalism and Development Research*. London: Pluto Press, 3–25.

Von Holdt, K. and Webster, E. 2008. 'Organising on the Periphery: New Sources of Power in the South African Workplace', *Employee Relations* 30 (4): 333–54.

Wacquant, L. 2003. 'Towards a Dictatorship over the Poor? Notes on the Penalization of Poverty in Brazil', *Punishment and Society* 5 (2): 197–205.

Wacquant, L. 2010. *Punishing the Poor: The Neoliberal Government of Social Insecurity*. Durham and London: Duke University Press.

Walder, A. G. (ed.). 1995. *The Waning of the Communist State: Economic Origins of Political Decline in China and Hungary*. Berkeley, CA: University of California Press.

Walters, J. and Cloete, D. 2008. 'The South African Experience with Negotiated versus Competitively Tendered Bus Contracts', *Transportation Research Part A: Policy and Practice* 42: 1163–75.

waMungai, M. and Samper, D. 2006. 'No Mercy, no Remorse: Personal Experience Narratives about Public Passenger Transportation in Nairobi, Kenya', *Africa Today* 52 (3): 51–81.

Wangwe, S. M. 1997. 'Small-Scale Mining and Mineral Stone/Gemstone Cross-Border Trade and Marketing in Tanzania'. Discussion Paper. Dar es Salaam, Tanzania: Economic and Social Research Foundation.

Wangwe, S. M. 2004. 'The Politics of Autonomy and Sovereignty: Tanzania's Aid Relationship', in S. Bromley, M. Mackintosh, W. Brown, and M. Wuyts (eds), *Making the International Economic Interdependence and Political Order*. Milton Keynes: The Open University, 370–412.

Watts, M. J. 2005. 'Baudelaire over Berea, Simmel over Sandton?', *Public Culture* 17 (1): 181–92.

WEAZ, Workers' Education Association of Zambia. 2006. 'Organising Informal Transport Workers: A Case Study of Zambia'. Research Report, June 2006. International Transport Workers' Federation.

Webster, E. 2015. 'Labour after Globalisation: Old and New Sources of Power'. ISER Working Paper No. 2015/1. Grahamstown, South Africa: Rhodes University, Institute of Social and Economic Research.

Weinstock, A. and Mutta, D. 2009. 'Bajajis Come to Tanzania', *Sustainable Transport*, Winter (21): 10–11.

Wells, J. 2001. 'Construction and Capital Formation in Less Developed Economies: Unravelling the Informal Sector in an African City', *Construction Management and Economics* 19 (3): 267–74.

Wilbur Smith Associates. 1991. 'Dar es Salaam Urban Passenger Transport Study'. Unpublished consultancy.

Wood, A. 2014. 'Moving Policy: Global and Local Characters Circulating Bus Rapid Transit through South African Cities', *Urban Geography* 35 (8): 1238–54.

Wood, A. 2015. 'The Politics of Policy Circulation: Unpacking the Relationship between South African and South American Cities in the Adoption of Bus Rapid Transit', *Antipode* 47 (4): 1062–79.

Wood, G. 2003. 'Staying Secure, Staying Poor: The "Faustian Bargain"', *World Development* 31 (3): 455–71.

Worger, W. 1983. 'Workers as Criminals: The Rule of Law in Early Kimberley, 1870–1885', in F. Cooper (ed.), *Struggle for the City: Migrant Labour, Capital and the State in Urban Africa*. London: Sage Publications, 51–90.

World Bank. 1989. 'Sub-Saharan Africa: From Crisis to Sustainable Growth'. Washington, DC: World Bank.

World Bank. 2009. 'Colombia Integrated Mass Transit Systems (IMTS): Additional Financing'. Project Information Document, Concept Stage Report No. AB4561. Washington, DC: World Bank.

World Bank. 2012. 'World Development Report 2013: Jobs'. Washington, DC: World Bank.

WRI, World Resource Institute. 2012. 'Statement: Development Banks Announce "Game Changer" for Sustainable Transport at Rio+20'. Washington, DC: World Resource Institute.

Wright, E. O. 2000. *Class Counts*. Cambridge: Cambridge University Press.

Wright, L. and Hook, W. 2007. 'Bus Rapid Transit Planning Guide'. New York: Institute for Transportation and Development Policy. Available at <https://www.itdp.org/wp-content/uploads/2014/07/52.-Bus-Rapid-Transit-Guide-PartIntro-2007-09.pdf>, accessed 1 February 2017.

Wuyts, M. 2001. 'Informal Economy, Wage Goods and Accumulation under Structural Adjustment: Theoretical Reflections Based on the Tanzanian Experience', *Cambridge Journal of Economics* 25 (3): 417–38.

Wuyts, M. 2011. 'The Working Poor: A Macro Perspective'. Valedictory address as Professor of Applied Quantitative Economics, 8 December. The Hague, Netherlands: Institute of Social Studies.

Wuyts, M. and Kilama, B. 2014. 'The Changing Economy of Tanzania: Patterns of Accumulation and Structural Change'. REPOA Working Paper 14/3. Dar es Salaam, Tanzania: REPOA.

Zeilig, L. and Ceruti, C. 2007. 'Slums, Resistance and the African Working Class', *International Socialism Journal* 117. Available at <http://isj.org.uk/slums-resistance-and-the-african-working-class/>, accessed 1 February 2017.

Interviews

All interviews were carried out in Dar es Salaam by the author.
Abasi. 2014. 19 February. Former *daladala* conductor and driver.
Adamu. 2014. 19 February. Beggar.
Airi. 1998. 18 December. Former *daladala* conductor.
Asenga. 2014. 19 February. Former *daladala* conductor.
Dongo, Issa. 2010. 3 August. Former *daladala* conductor.
Dongo, Issa. 2014. 17 February. Former *daladala* conductor.
Dotto. 2010. 12 August. Member of the route X workers' association.
Dotto. 2014. 18 February. Former *mpiga debe* and *daladala* worker.
Gwao. 2009. 16 September. General Secretary of DARCOBOA.
Kajembe. 2014. 21 February. *Daladala* driver.
Kizito. 2009. 4 September. Former *daladala* driver.
Kudo Boy. 2014. 18 February. *Daladala* driver.
Kulwa. 2009. 17 September. *Daladala* driver.
Mabrouk, Sabri. 2011. 21 December. DARCOBOA Chairman.
Mabrouk, Sabri. 2014. 13 February. DARCOBOA Chairman.
Mashaka. 1998. 10 December. *Daladala* driver on the bench.
Masuka, Juma. 2014. 18 February. *Daladala* driver.
Mayao, Martin. 2010. 3 October. Former *daladala* owner.
Mhina, Tambo. 2010. 9 August. Former DRTLA Secretary (1999–2005).
Mlaki. 1998. 15 November. Operations manager UDA.
Mlawa, S. 2014. 11 February. UWAMADAR General Secretary.
Mlawa, S. and Mnkeni, J. 2009. 10 September. UWAMADAR General Secretary and Treasurer, respectively.
Mtawali, Charles. 2014. 21 February. Tanzanian car importer.
Mudi Abasi. 2009. 17 September. *Daladala* conductor.
Muhidini. 2014. 19 February. Former *daladala* driver.
Mwaibula. 2010. Former DRTLA Chairman.
Natty. 2011. 15 December. Acting Director of Dares Salaam City Council.
Ndaombwa, George. 1998. 20 December. Chairman of MUWADA.
Ngaika, Rashidi. 2009. 16 September. *Daladala* driver.
Rajabu. 2014. 17 February. Taxi driver.
Rama. 2014. 19 February. Lorry driver.
Ramadhan. 1998. 8 December. *Daladala* driver.
Schelling, Dieter. 2011. 20 December. DART Chief Technical Adviser and former World Bank lead transport specialist in Tanzania.
Schelling, Dieter. 2014. 13 February. DART Chief Technical Adviser and former World Bank lead transport specialist in Tanzania.
Semvua, B. 2009. 10 September. COTWUT Deputy General Secretary.

Semvua, B. 2011. 5 December. COTWUT Deputy General Secretary.

Sulemani, L. 2009. 17 September. *Daladala* driver.

Sulemani, A. 2010. 8 September. Former Director of SUMATRA.

Sykes, Kleist. 2011. 7 December. Former Mayor of Dar es Salaam.

Thomas. 2014. 19 February. Former *daladala* conductor.

Tolu. 2010. 10 August. Former *daladala* conductor.

Toma. 2009. 17 September. Former *daladala* conductor.

Ubungo Station Manager. 2011. 14 December.

Uwazi. 2014. 19 February. *Daladala* conductor.

Index

Tables and figures are indicated by an italic *t* and *f* following the page number.

Abasi (former *daladala* conductor and driver) 88, 128, 134–5, 139, 141
accidents:
  bajaj and *bodaboda* 40
  daladala 2, 7, 23, 42, 72, 88, 91, 130
Adamu (beggar) 138
Africa:
  BRT systems 143
  infrequency of labour force surveys 57
  Institute for Transportation and Development Policy 148
  labour oversupply 67
  'maladjusted' states 49
  organizing informal labour 104
  second-hand Japanese cars 37–8
  transition to neoliberalism 27
  'unemployment spring' 157
  workforce criminalization 82
  see also cities, African; informal economy, African
'Africa rising' 19
agency:
  interplay with structure 12–13, 80, 103, 104, 122, 125, 140, 173, 175
  and optimistic informal economy narratives 2, 7–10, 12, 139–40, 175
  and optimistic urban narratives 4–7, 12, 139–40, 174
  overlooked in dystopic urban narratives 3, 122, 174
  and struggles for rights at work 101
ajira (employment) 77, 185
alcohol 132, 132 n.10
Alexander, P. 6
Amsden, A. H. 10
Arusha Declaration (1967) 28
Asenga (former *daladala* conductor) 71, 135, 136–8, 137*f*, 141
Asia 3, 24, 148
Askew, K. M. 18
'associational power' 86, 100, 101, 106, 119
Aziz, Rostam 147

Bagachwa, M. S. D. 30, 55
bajaj (three-wheeler auto-rickshaw) 38–40, 185
Benton, A. 175
Bernstein, H. 14, 82
bodaboda (two-wheeler moto-taxi) 38, 40, 185
Breman, J. 11, 14, 55, 73–4, 96, 103
bribes:
  to jail personnel 91
  to traffic police 47, 91, 93–4 n.13
British Bywater 151
Bryceson, D. F. 56
Bus Rapid Transit systems (BRTs):
  Centre of Excellence 148
  common characteristics of 143–4
  fares 149
  negative consequences of 149
  political economy of 26, 142–7
  as public–private partnerships 25, 143–4, 149, 151, 170, 178
  Tanzania (2002–16) 25–6, 142–3, 144–7, 152–70
  tensions as 'actually existing neoliberalism' 170
  vested interests of BRT evangelical society 26, 145, 148–9, 170, 173
  'win-win' narrative 26, 144, 146, 156, 157, 163, 170, 173
  see also Dar Rapid Transit (DART); Transmilenio, Bogotá

Calcutta rickshaw runners 73–4
Cape Town BRT 143, 153, 160
capital:
  and informal labour 23, 51–80
  mobility and vulnerability 103
  state alignment to, in informalization 11, 175
capitalism 9
  and class 13
  and economic informalization 11
  exclusion from as driver of informal economy 8, 8 n.6
  and labour oversupply 67

capitalism (*cont.*)
  urban 179
  see also class and socio-economic
    stratification; *daladala* owners; *daladala*
    workers; wage labour
car ownership, Dar es Salaam 27, 37–8, 152
  and traffic congestion 152–3
Carlyle Group 148
caste, and rigid labour market segmentation,
    India 96
Castells, M. 10–11, 102, 175
Central Transport Licensing Authority
    (CTLA) 35, 46, 47, 116
Chama cha Demokrasia na Maendeleo
    (CHADEMA) 155
Chama cha Mapinduzi (CCM) 18, 18 n.10, 19
  electoral motivation 146–7
  Magufuli's selection as presidential
    candidate 165
China 18, 39, 163
cities:
  and debates over neoliberalism 16
  demise of state-owned public transport 2
cities, African 3–7, 173–4
  'cardolatry' 27
  dystopic (structure-focused) narratives 2,
    3–4, 12, 122, 174
  effects of privatization and deregulation of
    public transport 42
  employment situation 3
  optimistic (agency-focused, populist)
    narratives 2, 4–7, 12, 84, 90, 139–40, 174
City Commission, Dar es Salaam 46
City Water, Dar es Salaam 151
class and socio-economic stratification 13–14
  daladala owner/worker divide 13, 23, 52–3,
    62, 75, 78, 100
  downplayed by de Soto 9
  and informalization 10–11, 175
  and regulation of daily life, Soweto 6
  as a relational concept 13–14
  and workers' power 85
  see also 'struggle over class'
'classes of labour' 14
  daladala workers 24, 84, 87–90, 90–1, 100
Cold War 30
collective bargaining 112–14
Communication and Transport Workers Union
    of Tanzania (COTWUT) 25, 75, 83,
    105 n.2, 173
  Coalition collective bargaining 112–14
  Coalition construction of shared meaning of
    exploitation 108–12
  Coalition demise 118–19
  Coalition efforts to strengthen state
    ties 116–18
  Coalition, general lessons from 119–21

  Coalition with UWAMADAR 107–8,
    119, 173
  correspondence with UWAMADAR and
    DARCOBOA 104
  engagement with 'informals' 107
  tensions with UWAMADAR over
    employment contracts 114–15
  visits exchanged with *daladala*
    workers 105–6
corruption 21 n.14, 82 n.2, 165
  and CCM 146–7, 147 n.6, 165
  and DART 154, 167–8
  and public–private partnerships 150–1

daladala (private buses) 1–2, 1 n.1, 22, 152,
    171, 185
  accidents 2, 7, 23, 42, 72, 88, 91, 130
  age of vehicles 40–1, 49, 153
  banning (1975) 30
  costs of purchase and operation 40
  divergence from specified routes 43
  early government toleration 30
  employment relations 13, 23, 47, 53, 61–3,
    71–6, 78–9, 80, 81, 112, 172
  identified as sole culprits of transport
    problems 153
  imprecise monitoring 46–7
  improvement possibilities ignored 153
  initial regulation (1983) 31–2
  market competition effects 1–2, 41–5
  names 42, 42–3 n.5
  numbers (1983–91) 33*t*
  numbers (1983–2010) 37*f*
  numbers, fluctuations (2000s) 27, 36
  numbers, increase in (1991–98) 34, 35, 36*t*
  overloading 1, 12, 13, 23, 42, 43, 73
  owners *see daladala* owners
  parallel operation with UDA-RT 164
  'pirate' 45–7
  'railway line' 91, 91 n.9, 135–6
  routes 28*f*
  size (2008–9) 49, 49*t*
  size of workforce 61–2 n.8
  speeding 12, 13, 23, 42–3, 42–3 n.5, 72, 73,
    82 n.2
daladala fares:
  1983–2013 35*f*
  and currency devaluation trends
    (1983–91) 32–3, 33*t*
  and currency devaluation trends
    (1991–2009) 34, 34*t*
  increasing marketization (1990–97) 34–5
  regulation attempt by SUMATRA 48
  state regulation (1983–90) 32–3
  trends in school pupil and adult 43*t*
  wars following liberalization 41
  workers' co-operation over 91

daladala owners 63–6
  class divide with workers 13, 23, 52–3,
    62, 71–6, 78, 100
  and DART 157, 158–61, 162–3, 178
  differentiation among 66
  difficulties of incorporation into
    DART 158–61
  vs. 'name lenders' 63
  number of buses owned 33, 47, 63
  power asymmetry with workers 23, 73, 75,
    80, 83, 85, 93, 117, 172
  profitability of the business 64–5
  'relay' vs. 'accumulation' groups 65–6
  as rentiers 74
  responsibility for violations of road safety
    rules 116, 117
  socio-economic background 64
daladala owners' association *see* Dar es Salaam
  Commuter Bus Owners Association
  (DARCOBOA)
daladala workers 66–71
  age and aging 73, 73 n.17, 131
  age and occupational immobility 140, 173
  aspirations for work progression 125
  'associational power', lack of 86, 100, 106
  attitude towards *bajajis* and *bodaboda* 40
  barriers to collective organization 24, 86–90,
    100, 172
  class divide with owners 13, 23, 52–3, 62,
    71–6, 78, 100
  'classes of labour' 24, 84, 87–90, 90–1, 100
  criminalization 23, 24, 81–2, 98, 110, 172
  DART implications for 157–8
  disillusionment with government 82
  educational level 68, 115
  employment contract demands 75, 109,
    110, 112, 173
  employment contract enforcement 114–15,
    117–18
  employment contract statistics (1998, 2003)
    71
  employment histories 68–71
  fake driving licenses 68, 68 n.11
  'forever' 125
  health problems 128, 134
  horizontal mobility and implications 84,
    96–8
  imprisonment 81, 91
  lack of respect for 128, 133
  length of working day 72–3
  'Life is War' 23, 52*f*, 73, 80
  'marketplace power', low 85, 98, 100, 113
  'no more' 125
  occupational mobility/immobility 25,
    122–41, 173
  'out and back in' 125
  oversupply 67–8, 80, 85, 86, 113, 140, 172

  payment modalities 127–8
  political mobilization (1997–2014) 25,
    105–21, 172–3, 179
  political quiescence (pre-1997) 24–5,
    81–99, 172
  power asymmetry with owners 23, 73, 75,
    80, 83, 85, 93, 117, 172
  as real living people 175–6
  relationship with school pupils 1, 13, 23,
    42–4, 47, 106
  rental to owners 72, 73, 82, 85, 109
  response to exploitation 13, 72–3, 172
  returns from work 71–2, 73–4, 104
  sex and access to jobs 62
  'slowly out' 125
  spatial unit of work 86
  'structural power' 112–13, 121
  'struggle over class' 24, 84, 90, 94–6, 98
  as 'temporary owners' 117
  transition from conductors to drivers 140,
    140 n.13
  wildcat strikes and localized walkouts 113
  working conditions 87–8, 108–9, 128, 175
  'workplace power' 85–6, 98, 100
  writings and drawings on buses 1, 3, 40–1, 52*f*,
    66–7, 67*f*, 72–3, 74*f*, 82, 122–3, 175, 176*f*
daladala workers association, formal *see*
  UWAMADAR
daladala workers association, informal
  (1998–2005) 24–5, 83–4, 90–6, 98–9, 100
  archive 97 n.20
  break-up 99
  'carrying capacity' 97
  grants and loans 93
  inclusion and exclusion dynamics 97–8
  managing precariousness 90–3, 172
  prevention of competition 91–2
  spatial advantages 91
  'struggle over class' 94–6, 98
  welfare fund 92–3, 98, 99
daladalaman maisha ('for life'/'with a
  livelihood') 87–8, 90–1, 92, 97–8, 129–30,
  131, 185
Dar es Salaam:
  culture of non-compliance 42–3 n.5
  dockworkers' strike (late 1940s) 86
  formalization of property rights 10
  government withdrawal from services 27
  population 2, 27, 30
  private car ownership 27, 37–8, 37 n.3, 152–3
  public sector workers' car allowance 38
  traffic congestion 38, 39*f*, 40, 143, 152–3
  water supply corruption scandal 151
  see also public transport
Dar es Salaam City Council (DCC) 46
  and DART 146, 152, 153–4, 155, 156, 159
  ownership of UDA 161

Dar es Salaam Commuter Bus Owners Association
   (DARCOBA) 63 n.9, 104, 112, 115
   and DART 145, 159–60, 163
   and employment contracts 112, 113–14,
      115, 118
   and UDA-RT 164
Dar es Salaam Motor Transport Company
   (DMT), nationalization 28–9
Dar es Salaam Transport Licensing Authority
   (DRTLA) 47–8, 159
   and *daladala* workers' employment
      contracts 115
Dar Rapid Transit (DART) 25–6, 143, 144–7,
   152–70, 173, 178
   compensation disputes 155
   corruption scandal 167–8
   and *daladala* owners 157, 158–61
   deeper roots of government lack of
      support 157–62
   delays in implementation 144, 153, 163
   delays in implementation rooted in
      resistance and politics 26, 144–6, 147,
      153–7, 163, 170, 173
   employment implications 157–8
   and fare increases 153, 164, 167, 168–9
   Gerezani dispute 156
   ideology 152–3
   operational debut (2016) 169
   political economy 145
   and President Magufuli 166–70
   towards implementation (2014
      onwards) 162–5
   Ubungo station dispute 154–5
   and UDA 161–2
DART *see* Dar Rapid Transit
Davis, M. 3–4, 4 n.3, 7, 86–7, 95, 97, 102,
   174 n.1
Day of the African Child (1991) 44
day waka/waka (*daladala* worker 'on the
   bench') 69, 72, 87–8, 90, 91, 92, 97, 98,
   129, 130, 131, 132, 133, 136, 139, 185
De Boeck, F. 5 n.4
De Soto, H. 8–10, 11, 12, 54, 174, 178
deregulation:
   from below 10
   public transport 16, 22, 31–7, 49, 142, 171,
      177–8
   public transport effects 12, 13, 23, 41–5, 49,
      73, 80, 81, 93, 171, 172, 178
   and 'roll-back' neoliberalism 15
dereva (driver) 185
developmentalism:
   failure 3, 29, 174
   transition to neoliberalism 3, 27, 49
Dongo (former *daladala* conductor) 77, 85,
   87–8, 125, 135–6
Doogan, K. 103

Dotto (former *daladala* worker) 89, 127,
   135–6, 141
Due, J. M. 45

East African, The 46
economic crisis (1970s) 17
economic growth and job creation 19–21
EMBARQ (World Resource's Institute for
   Sustainable Cities) 148
employment contracts:
   barriers to enforcement 114–16
   collective bargaining for 112
   daladala statistics (1998, 2003) 71
   daladala workers' demands 75, 109, 110,
      112, 173
   DARCOBOA's ambiguous response 113–14
   formal sector 54
   state role in enforcing 117–18
employment (labour) relations, informal:
   as barrier to workplace labourism 101–2,
      112, 120–1
   consequences of unregulated 13, 23, 73, 80,
      81, 82, 93, 172
   daladala 13, 23, 47, 53, 61–3, 71–6, 80, 81,
      112, 172
   formalization through workplace
      labourism 103, 105, 111, 112, 114, 121
   need to tackle unregulated 178
   Transmilenio 149
employment rights *see* labour/employment
   rights
Engels, F. 15

Fields, G. S. 55 n.5
Fischer, G. 104
Fisher, G. 110
Fold, N. 56
formalization of property rights 10, 51, 178
Fox, L. 51
free markets 12, 15, 171, 176
Friedman, Milton 15
functionalism 2, 6, 120
   avoiding 11, 175

Gallin, D. 102
'generative powers' 5, 90
Gerezani, Dar es Salaam 156
globalization, and labour possibilities 25, 101, 103
Goldman Sachs Urban Investment Group 148
Gooptu, N. 14, 24, 54, 96

Harriss-White, B. 14, 24, 54, 96
Hart, G. 17
Hart, K. 54
Hayek, Friedrich 15
hesabu (daily rent) 71, 108, 128, 135, 185
horizontal mobility 84, 96–8

India, informal economy 11, 96
  'struggle over class' 14
industrial revolution, Britain 67
informal economy:
  capital and labour 23, 51–80
  economic mobility myth 174
  ILFS (2006) statistics 53, 57–61, 58*t*, 59*t*, 60*t*,
    76–8, 79
  India 11, 14, 96
  and labour fragmentation 86–7
  and labour market dualism 55 n.5
  as 'last resort' 3, 55
  'misplaced aggregation' fallacy 54–5, 79
  petty exploitation as essence 95–6
  possibilities for collective action 25, 101–5,
    112, 120–1, 173, 175
  relationship between trade unions and
    workers 103–4, 107, 119–20
  self-employment seen as predominant 9–10,
    11, 23, 51–2, 52 n.3, 53–6, 57–9, 60, 78,
    172, 175, 178
  as 'training ground' 127, 140, 173, 174
  variety of production and labour regimes 14,
    54, 55–6, 79
  wage labour downplayed or invisible 11, 23,
    51, 51 n.1, 53, 55, 57, 59, 60–1, 75–6, 77–8,
    78–80, 175
  see also employment (labour) relations,
    informal
informal economy, African 7–12
  dystopic (structure-based) narratives 2,
    10–12, 175
  optimistic (agency-based) narratives 2, 7–10,
    12, 139–40, 175
  size 7, 7 n.5
informal transport workers *see daladala* workers
informalization 2, 171
  over-regulation seen as driver 9
  and rights at work 100–5
  role of class and structural forces 10–11, 175
Institute for Transportation and Development
  Policy (ITDP) 144, 148, 152
Integrated Labour Force Survey (ILFS, 2006)
  53, 79
  definitions and patterns of
    employment 57–61, 58*t*, 59*t*, 60*t*
  translation issues 53, 76–8
International Monetary Fund (IMF) 19, 31, 34
Investment Promotion and Protection Act
  (1990) 34

Jamal, V. 55
Japanese second-hand cars 37–8
jobless growth 19–20, 67–8
'jobs dementia' 10
Johannesburg 4, 5, 6
  BRTs 143, 153, 160

Jonssøn, J. B. 56
Jumbe, Mr 39

Kajembe (*daladala* driver) 69, 131, 134
Kampuni ya Mabasi ya Taifa (KAMATA) 29
Kanyama, A. 64–5
Kelsall, T. 146–7
Kenya 42
kibarua (work without contract) 62, 77, 79, 185
Kibasila Society Group 156
kijiwe (pavement) 95, 95 n.16, 130, 185
Kikwete, Jakaya 157–8, 159, 161–2
Kinshasa 5 n.4
Kisena, Robert 161–2, 162 n.21, 168
kondakta/konda (conductor) 185
Kudo Boy (*daladala* driver) 69
kukomaa (overripen) 87, 185
Kulwa (*daladala* driver) 139
Kushoka, Mr 34
kutafuta maisha (try to build your life) 69, 185

labour 'classes' *see* 'classes of labour'
labour force surveys 56, 57, 80
  design 54
  ILFS 53, 57–60, 58*t*, 59*t*, 60*t*, 76–8, 79
  problems generated by OECD origins 60–1
  self-employment seen as predominant 51–2,
    57–9, 60, 61
  translation issues 76–8, 79, 172
  wage labour/paid employment ignored or
    downplayed 53, 60–1, 75–6, 77–8, 172
labour fragmentation 14, 84, 86–7
  as barrier to organization 14, 24, 86,
    96, 98
  and workers identity 24, 90, 94
  see also 'classes of labour'
labour heterogeneity 14, 86–90
labour market dualism 55
labour oversupply:
  as barrier to collective action 14, 24, 98
  and exploitation 172
  and low 'marketplace power' 85
  and workers' identity 24, 84, 90, 94
labour politics:
  quiescent period (up to 1997) 81–99
  struggling for rights at work
    (1997–2014) 100–21
labour/employment rights:
  barriers to enforcement 114–15
  bringing the state back in 115–18
  through collective bargaining 112–14
  as early goal of *daladala* workers 106
  and informalization 100–5
labour unrest, ongoing global 103, 112
Latin America:
  BRT alleged successes 144
  economic informalization 10–11

Latin America: (*cont.*)
  ITDP 148
  see also Transmilenio, Bogotá
Lerche, J. 14
Lowassa, Edward 147, 147 n.6, 154, 155
Lwatakare, Ronald 168

Mabruk, Sabri 166–7
Magufuli, John 155, 165–7
maisha (life/livelihood) 87, 185
Makamba, Lieutenant 116
Maliyamkono, T. L. 30, 55
Mamdani, M. 14
Mamuya, I. 30
market fundamentalism 9, 12, 175
'marketplace' power 85, 98, 100, 113
Marx, K. 9, 13, 15, 67, 72
Marxist approaches 5–6, 13, 14
Masuka, Juma (*daladala* driver) 68–9, 129–30
Mbembe, A. 6
Meagher, K. 11
methodology 21–2
  challenges of studying bus owners 66
  issues in studying occupational mobility/
    immobility 122–9
Mfinanga, D. 161
Mhina, Tambo 47, 63, 159
microcredit/microfinance 10, 51, 75, 178
Minister of Home Affairs, Tanzania 44, 45, 81
Ministry of Communication and Transport,
    Tanzania 31, 32, 33, 35, 45, 63
Ministry of Work, Employment and
    Development, Tanzania 117
Mkapa, Benjamin 147
mkataba (contract) 185
MKURABITA (Property and Business
    Formalization Programme), Tanzania 10
mkuu wa reli (leader of *daladala* workers'
    'railway line') 135, 185
Mlambo, Asteria 167–8
Mnyika, John 155
Monson, Jamie 18
mshahara (wage) 77, 185
Mtawali, Charles 38
Mudi Abasi (*daladala* conductor) 138–9
Muhimbili Hospital, Dar es Salaam
    165–6
Mushi, Mr 163
Mwinyi, Ali Hassan 44
Myers, G. 3–4

Nairobi 42
National Road Safety Policy 152
National Transport Corporation 31
nationalization 28–9
Natty, Mr 156
neoliberalism 14–17
  'actually existing' 16, 26, 28, 142, 144–5,
    170, 172, 178
  African workers' political mobilization
    under 101
  debate on usefulness of term 14, 16–17,
    176–7
  emergence of concept 15
  grounding 17, 176–9
  ideological dimension 15, 177, 178
  new face in Tanzania 142–70
  pernicious consequences 3
  as political project 15, 177
  and post-socialism 17–21
  promotion in Tanzania 17, 22, 34
  'roll back'/'shock therapy' 15–16, 142, 171,
    177–8
  'roll-out' 15–16, 143, 177, 178
  'spontaneous combustion' 45, 142
  and state intervention 15–16
  transition from developmentalism to
    27, 49
newspapers, as research sources 22, 104
Ngaika, Rashidi (*daladala* driver) 77, 131
Nuttall, S. 6
Nyanda, Frenky (taxi driver) 135
Nyere, Julius 19

occupational mobility/immobility 25,
    122–41, 173
  histories of occupational immobility 129–32
  histories of occupational mobility 132–9
  methodological issues 122–9
  workers trajectories 139–41
oil crisis (1970s) 11
'ordinary, the' 4–5, 6, 84
Organization for Economic Co-operation and
    Development (OECD) 51, 60, 61

paid employment *see* wage labour
Pakistani car traders 38
Parnell, S. 177
'people as infrastructure' 5, 6
Peru 9
Pieterse, E. 4, 5, 7, 8, 8–9 n.7, 12, 174
Pikine, Senegal, informal economy 7–8
Pimhidzai, O. 51
Pirie, Gordon 145
Pitcher, A. M. 18
Planet of Slums (Davis) 3–4, 4 n.3
Plissart, M. F. 5 n.4
political economy 4, 17, 21, 101
  benefits as analytical approach 120
  blindness to 104
  of BRTs 26, 142–7, 173
  of Japanese vehicle production 37
  of labour relations 103
  Marxist 5–6, 13

Portes, A. 10–11, 102, 175
postcolonial state development *see*
     developmentalism
postcolonial urban narratives 4–5, 12
post-socialism 17–22, 31–3, 147
poverty:
  blamed on the poor 82
  eradication fantasies 3, 174
  slow reduction during economic growth,
       Tanzania 19–20, 67
  structural causes 5
power, sources of workers 84–6
power balance:
  daladala owners and workers 23, 73, 75, 80,
       83, 85, 117, 172
  labour/capital 103
Prime Minister's Office, Tanzania 143,
     154, 163
private car ownership, Dar es Salaam 27,
     37–8, 152
  and traffic congestion 152–3
privatization:
  public transport 2, 16, 22, 27, 31–7, 49, 142,
       171, 177–8
  public transport effects 41–5, 49, 171–2
  resistance to TAZARA 15–16
  and 'roll-back' neoliberalism 15
  UDA, Dar es Salaam 161–2
  unsuccessful as panacea for economic
       growth 45
  see also public–private partnerships
property rights, formalization 10, 51, 178
public–private partnerships (PPPs) 16, 149–51
  Bus Rapid Transport projects as 25, 143–4,
       149, 151, 170
  economic case for 150
  justification dubious 150–1
  weak bargaining power of public regulator
       bodies 149
public transport 22–3, 27–49
  bajaj and *bodaboda* 38–40
  capital and informal labour 23, 51–80
  changing nature of social conflict 44 n.7
  control by organized crime 42
  declining supply (1970–83) 29–30
  demise of state-owned 2
  feeble attempts to regain public control
       (1999–2015) 23, 45–49, 142
  as 'functional chaos' 1–2, 12
  grounding neoliberalism 17, 177–9
  neoliberalism, post-socialism and 14–21
  privatization and deregulation, effects on
       quality 12, 13, 23, 41–5, 49, 73, 80, 81, 93,
       171–2, 178
  privatization and progressive deregulation
       (1983–2001) 2, 16, 22–3, 27, 31–7, 49,
       142, 171, 177–8

public debt incurred to overhaul 16, 178
  re-organization under public–private
       partnership (2002–16) 16, 142–71
  research methodology 21–2
  rising demand 30–1
  state-provided (1970–83) 2, 28–31
  and 'workplace power' 85–6, 98, 100
  see also Bus Rapid Transit systems; *daladala*;
       Dar Rapid Transit

Quinn, S. 55

Rajabu (taxi driver) 70, 128, 133, 141
Rama (lorry driver) 70, 127, 128
Reagan, Ronald 15
religion:
  and *daladala* division of labour 88 n.5
  and rigid labour market segmentation,
       India 96
Robinson, J. 4, 12, 174, 177

Sarris, A. H. 55
school pupils 1, 13, 23, 42–4, 47, 106
Sea and Maritime Transport Regulating
     Authority (SUMATRA) 117
self-employment:
  combination with employment 14
  and *daladala* sector 53, 66, 74–5, 78–9
  labour force survey definition 57
  perceived dominance in the informal
       economy 9–10, 11, 23, 51–2, 53, 53–6,
       57–9, 60, 78, 172, 175, 178
  translation biases towards 53, 78, 79, 172
'self-reliance' 19, 19 n.12
Semvua, B. 105 n.1, 114, 118
Senegal informal economy analysis 55–6
sex (gender) as access to *daladala* jobs 62
Shirika la Usafiri Dar es Salaam (UDA)
     29–31, 63
  and DART 161–2, 163
  demise (1974–98) 2, 29–30, 29t, 36, 161
  licensing powers 31–2, 33
  licensing powers removed 35
  privatization 36 n.2, 161–2
  tax issue with TRA 167
  and UDA-RT 164
Shivji, I. G. 107
Silver, B. J. 85–6, 103, 113, 121
Simba, Iddi 147, 161 n.20
Simon Group 161, 169–70
Simone, A. 4–5, 7–8, 8–9 n.7, 12, 174
social protection 102, 120, 120 n.23
South Africa:
  collective bargaining in the informal
       economy 112
  control of urban by organized crime 42
Soweto 4, 6

Standing, G. 101
state, the:
 alignment to the interests of capital in
  informalization 11, 175
 neoliberal 'rolling back' 15–16, 142, 171,
  177–8
 neoliberal 'rolling out' 15–16, 143, 177, 178
 over-regulation seen as driver of
  informalization 9
 structural adjustment programmes 3, 150
 Tanzania 22, 31, 32, 34, 38, 171, 178
structure:
 as barrier to political organization 101, 103
 in dystopic narratives of the African city 2,
  3–4, 12–13, 174
 interplay with agency 12–13, 80, 103, 104,
  122, 125, 140, 173, 175
 overlooked in optimistic narratives of the
  African city 6–7, 12, 174
 overlooked in optimistic narratives of the
  African informal economy 8, 9, 12,
  122, 175
 in pessimistic narratives of the African
  informal economy 10–12, 175
 'struggle over class' 14
 daladala workers 24, 84, 90, 94–6, 98
Sulemani (daladala driver) 131–2
Surface and Marine Transport Regulatory
  Authority (SUMATRA) 48–9, 63, 64,
  65, 118
 Consumer Consultative Council 167
Sykes, Kleist 152, 154, 158

Takule, Cosmas 154–5
Tanganyika African National Union
  (TANU) 18
 and trade unions 106–7
Tanzania:
 BRT project (2002–16) 25–6, 142–3, 144–7,
  152–70
 corruption in public–private
  partnerships 150–1
 DART, government involvement and control
  under Magufuli 166–70
 DART, government tepid commitment 146,
  147, 157–62, 170
 demise of developmentalism 29
 demographic trends 20, 20 n.13
 foreign investment 20–1, 34
 formalization of property rights 10
 GDP 19
 government attempt to tackle speeding
  82 n.2
 government decline of economic
  capacity 30–1
 government feeble re-regulatory efforts
  (1999–2015) 23, 45–9, 142, 172, 178

government limited capacity for
  intervention 44–5, 46–9, 142, 172
 government mixed commitment to
  economic liberalization 17–18, 22–3,
  31, 32
 ILFS (2006) 53, 57–61, 58t, 59t, 60t, 76–8, 79
 illegal immigrants 166
 independence (1961) 22, 107
 informal construction industry 56, 60
 informal economy growth potential 55
 informal mining industry 56, 60
 negative impact of privatization of vegetable
  oil industry 45
 neoliberalism, 'actually existing' 26, 28, 142,
  170, 172, 178
 neoliberalism, promotion 17, 34
 neoliberalism, transition from
  developmentalism to 27, 49
 policy changes (1990–1) 34–5
 policy reversals (2000 onwards) 21, 21 n.14
 policy towards school pupils and
  daladala 43–5
 political landscape, post-liberalization 146–7
 post-socialism 17–21, 31–3
 provision of public transport (1970–83)
  2, 28–31
 public debt incurred to overhaul public
  transport 16, 178
 regulation of private sector transport
  (1983–90) 31–3
 state role in enforcing labour rights 115–18
 structural adjustment programmes 22, 31,
  32, 34, 38, 171, 178
 trade unions 104, 105, 106–7, 110
 ujamaa period (1967–mid 1980s) 18–19,
  28–31
 urban poverty blamed on the poor 82
 World Bank survey experiment 60–1
Tanzania Drivers Association (TDA) 116
Tanzania Revenue Authority 46, 46 n.8
 standoff with UDA-RT 166–7
Tanzania Road Transport Workers Union
  (TARWOTU) 118
Tanzanian Red Cross 39
Tanzanian transport union see Communication
  and Transport Workers Union of Tanzania
  (COTWUT)
Tanzania–Zambia Railway (TAZARA) 18–19,
  18 n.11
Teal, F. 55
Temu, A. E 45
Thatcher government, UK 15, 150
Thomas (former daladala conductor) 128
Tolu (former daladala conductor) 126, 139
trade unions and the informal economy:
 dynamics of conflict in renewal 119
 pessimistic narratives 101–2

relationship with workers 103–4, 119–21
trade unions, Tanzania:
    attention to 'informals' 107
    legal rights of representation 105
    need for supporting majority 110
    political-economy-blind study (2013) 104
    relationship to the ruling party 106–7
    see also Communication and Transport
        Workers Union of Tanzania (COTWUT)
Traffic Police Department 31–2, 47
Transmilenio, Bogotá 144, 145, 148–9, 152
    negative consequences 149
Tripp, A. M. 10, 55
Twaweza (Tanzanian NGO) 167

Ubungo station 154–5
UDA *see* Shirika la Usafiri Dar es Salaam
UDA-RT *see* Usafiri Dar es Salaam Rapid Transit
ugali (Tanzanian diet staple) 67 n.10
ujamaa (Tanzania's socialist
    experiment) 18–19, 28–31, 186
UN Sustainable Development Conference, Rio
    de Janeiro 144
United Kingdom (UK):
    neoliberalism 15
    public–private partnerships 150
United States 15
urban capitalism 179
urbanization 16, 173–4
    and neoliberalism 16
    without economic growth and
        industrialization 3
Usafiri Dar es Salaam Rapid Transit (UDA-
    RT) 163–4
    alleged contract violation 168
    fares 167, 168–9
    standoff with Tanzania Revenue
        Authority 166–7
UWAMADAR (*Umoja wa madereva na
    makondakta wa mabasi ya abiria Dar es
    Salaam*, transport workers'
    organization) 24, 25, 62, 83, 100–1,
    105 n.2, 173
    Coalition collective bargaining 112–14
    Coalition's construction of shared meaning
        of exploitation 108–12
    Coalition correspondence with bus owners'
        association 104
    Coalition demise 118–19
    Coalition efforts to strengthen ties to the
        State 115–18

Coalition, general lessons from 119–21
Coalition with COTWUT 107–8, 119, 173
correspondence with COTWUT and
    DARCOBOA 104
early days (1995–2000) 105–8
establishment (2000) 75, 83, 107–8
recruitment challenges and strategy
    110–12
tensions with COTWUT over employment
    contracts 114–15
Uwazi (*daladala* conductor) 69, 130–1

Van den Brink, R. 55
vipanya (small minibuses) 49, 86, 186
Volvo 148

wage labour:
    'formal' definition 54
    hidden or overlooked in informal
        economy 11, 23, 51, 51 n.1, 53, 55, 57, 59,
        60–1, 75–6, 77–80, 175
    labour force surveys definition 61
    as main source of income to the poorest 14
    predominance in *daladala* sector 62,
        75–6
    in Senegalese petty production 55–6
    translation issues 77–8, 79
Wangwe, S. M. 56
wapiga debe ('those who hit the tin') 89–90, 91,
    92, 135, 186
Watts, M. J. 5–6
watu wa benchi (people on the bench) *see day
    waka*
Weeks, J. 55
work, spatial unit of 86
workers' associationism 24–5
    ethnography 84
    forms and limits of solidarity 90–6
workers' power:
    contextualizing 119–21
    sources 84–6, 172
workforce criminalization 23, 24, 81–4, 98,
    110, 172
'workplace power' 85–6, 98, 100
World Bank 19, 21 n.14, 31, 34, 45, 51, 149,
    152–3
    and BRTs 143, 148, 149
    and DART 16, 25, 143, 146, 152, 160, 163,
        170, 178
    survey experiment 60–1
Wright, E. O. 84–5, 121

Printed and bound by CPI Group (UK) Ltd, Croydon, CR0 4YY